AF837509

Pulsed Electron–Electron Double Resonance

Yuri D. Tsvetkov · Michael K. Bowman
Yuri A. Grishin

Pulsed Electron–Electron Double Resonance

Nanoscale Distance Measurement in the Biological, Materials and Chemical Sciences

Springer

Yuri D. Tsvetkov
The Voevodsky Institute of Chemical Kinetics and Combustion
Novosibirsk, Russia

Yuri A. Grishin
The Voevodsky Institute of Chemical Kinetics and Combustion
Novosibirsk, Russia

Michael K. Bowman
Department of Chemistry and Biochemistry
University of Alabama
Tuscaloosa, AL, USA

ISBN 978-3-030-05371-0 ISBN 978-3-030-05372-7 (eBook)
https://doi.org/10.1007/978-3-030-05372-7

Library of Congress Control Number: 2018963988

© Springer Nature Switzerland AG 2019
This work is subject to copyright. All rights are reserved by the Publisher, whether the whole or part of the material is concerned, specifically the rights of translation, reprinting, reuse of illustrations, recitation, broadcasting, reproduction on microfilms or in any other physical way, and transmission or information storage and retrieval, electronic adaptation, computer software, or by similar or dissimilar methodology now known or hereafter developed.
The use of general descriptive names, registered names, trademarks, service marks, etc. in this publication does not imply, even in the absence of a specific statement, that such names are exempt from the relevant protective laws and regulations and therefore free for general use.
The publisher, the authors and the editors are safe to assume that the advice and information in this book are believed to be true and accurate at the date of publication. Neither the publisher nor the authors or the editors give a warranty, express or implied, with respect to the material contained herein or for any errors or omissions that may have been made. The publisher remains neutral with regard to jurisdictional claims in published maps and institutional affiliations.

This Springer imprint is published by the registered company Springer Nature Switzerland AG
The registered company address is: Gewerbestrasse 11, 6330 Cham, Switzerland

Preface

Following the discovery of electron paramagnetic resonance (EPR) by E. K. Zavoisky in the Russian city of Kazan more than 70 years ago, applications of EPR for studies of the structure and properties of paramagnetic particles and materials have flourished. This is due, in part, to the technical development and growth of the EPR method itself. Starting from the 1960s, pulse EPR has overtaken continuous-wave EPR methods. This occurred through the development of electron spin echo, multipulse and two-dimensional sequences; double electron–nuclear and electron–electron resonance techniques; triple resonance; EPR tomography; and high-frequency EPR methods extending into the millimeter wavelength range. All these directions of EPR development are based on its unique combination of high sensitivity, selectivity, and resolution in both the frequency and time domains. New horizons continue to arise for applications in physics, chemistry, and biology. Traditionally, pulse EPR methods and their numerous variants are most widely used in physicochemical applications, such as the structure and properties of atoms, ions, radicals, and molecules; chemical kinetics; and physical dynamics.

In this book, we have tried to summarize the development and application of a relatively novel variant of EPR spectroscopy, i.e., the method of pulsed electron–electron double resonance (PELDOR) or double electron–electron resonance (DEER); we will use the first abbreviation in this book. PELDOR was proposed and developed in 1981 in Novosibirsk in the Voevodsky Institute of Chemical Kinetics and Combustion by A. D. Milov, K. M. Salikhov, and M. D. Schirov.

PELDOR is the offspring of other pulse EPR methods, particularly electron spin echo (ESE), in which all of the authors worked. One prominent ESE application in chemistry is to the study of dipole–dipole magnetic interactions between paramagnetic centers. Such work provides information about the spatial distribution and distances between spins in disordered systems. However, this information is hard to extract because the analysis of ESE data about dipole–dipole interactions is complicated by other interactions within the spin system, i.e., electron–nuclear interactions and relaxation processes such as spectral diffusion. The potential of ESE is also restricted by the so-called dead time, which limits the measurement of short relaxation times.

PELDOR overcomes many of these difficulties. It became possible: to measure the distances and the distribution of distances between pairs of spins for biradicals and radical pairs; to determine the geometry spin-labeled molecules; to estimate the number of interacting spins in spin clusters; and to characterize the spatial distribution of paramagnetic centers. The limitations, caused by instrumental dead time which limited ESE to distances greater than 1.5 nm, are absent in PELDOR.

As a result, PELDOR is now a fairly popular method in EPR spectroscopy, particularly as applied to biologically important systems. PELDOR is becoming more readily accessible to researchers due to the appearance of PELDOR spectrometers in a wide range of frequency bands.

This book covers the basic theory of PELDOR, the basic techniques, and some applications. Already, several hundred published articles concern applications of PELDOR in chemistry and biology. The choice of examples in our book is, of course, subjective and we cannot claim to completely cover all application of PELDOR. We do hope that the interested reader will come to an appreciation of the opportunities PELDOR provides and will find the key to solve their own physicochemical and biochemical problems.

The authors take pleasure in expressing their gratitude to colleagues in PELDOR over the years: A. D. Milov, A. G. Maryasov, S. A. Dzuba, and R. I. Samoilova, for their extensive experimental and theoretical collaborations; and to C. Toniolo, F. Formaggio, J. Raap, and O. S. Fedorova for their participation and valuable discussions in our research; and to O. V. Polukarikova for her assistance and help in preparation of this manuscript.

Novosibirsk, Russia — Yuri D. Tsvetkov
Tuscaloosa, USA — Michael K. Bowman
Novosibirsk, Russia — Yuri A. Grishin
May 2018

Contents

Chapter 1
Introduction

Abstract PELDOR or DEER spectroscopy is a magnetic resonance method to probe spin interactions between free radicals. It was proposed and demonstrated in the 1980s and has developed into a powerful means to routinely measure distances in non-crystalline chemical, macromolecular and biological systems on the nanometer length scale. The basis of PELDOR spectroscopy is introduced using a simple model with two unpaired electron spins.

1.1 PELDOR Pulse Sequence and Vector Model

The Pulsed Electron Double Resonance (PELDOR) technique was proposed in 1981 to probe spin interactions between free radicals [1, 2]. Modulation of the PELDOR signal due to spin-spin interactions was observed in 1984 [3]. Since its introduction, PELDOR has been used to study magnetic interactions and distances in a wide variety of chemical, macromolecular and biological systems. It has become a powerful means of quantitatively characterizing structures and structural changes on the nanometer length scale. The PELDOR experiment and resulting spectrum are readily understood from a simple model with two unpaired electron spins.

1.1.1 Three-Pulse PELDOR

The original pulse sequence, known as 3pPELDOR, consisted of three microwave (mw) pulses at two different frequencies, Fig. 1.1. These pulses were applied to pairs of free radicals with spin $S = 1/2$, often referred to as spin labels. The mutual spin-spin interaction D between the spin labels is the combined magnetic dipole-dipole ω_D and exchange J interactions:

© Springer Nature Switzerland AG 2019

Y. D. Tsvetkov et al., *Pulsed Electron–Electron Double Resonance*,
https://doi.org/10.1007/978-3-030-05372-7_1

$$D = \omega_D\left(1 - 3\cos^2\theta\right) + J \tag{1.1}$$

$$\omega_D = 2\pi\,\nu_D = \frac{\hbar\gamma^2}{r^3} = \frac{327.7\,\text{MHz}\,\text{nm}^{-3}}{r^3} \tag{1.2}$$

where θ is the angle between the direction of the magnetic field $\vec{B}_0$ and the vector $\vec{r}$ connecting the centers of the unpaired electrons, Fig. 1.1c; and γ is the gyromagnetic ratio for a free electron. If the frequency is in MHz, the dipolar distance r in nm between electron spins is $r = [52.16/\nu_D]^{1/3} = [327.7/\omega_D]^{1/3}$.

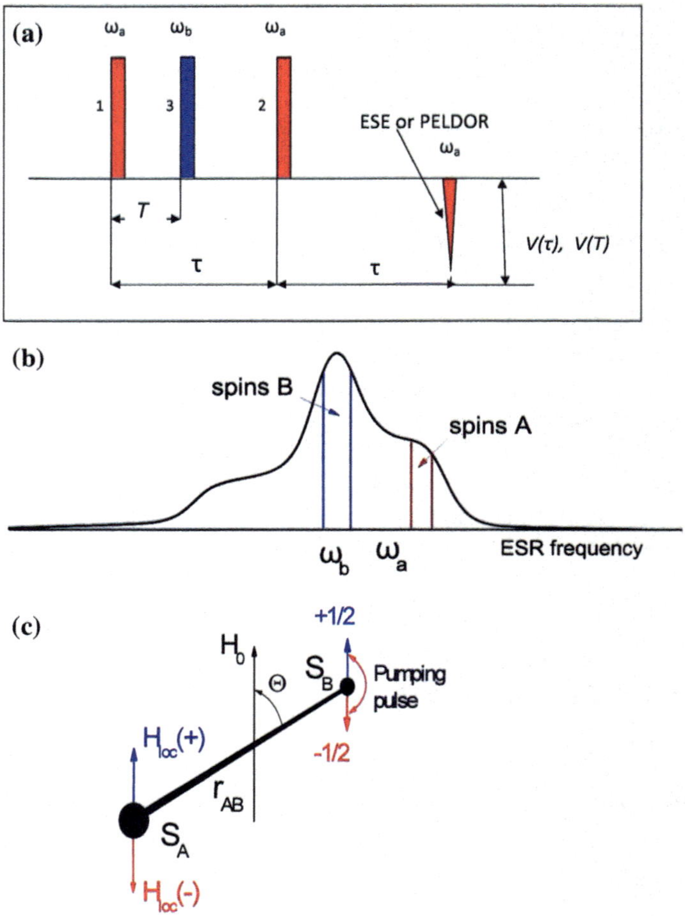

Fig. 1.1 The 3pPELDOR technique: **a** observe pulses at ω_A (red) at times 0 and τ produce a two-pulse spin echo from A spins and a pump pulse at ω_B (blue) changes its ampllitude; **b** the A and B spins have EPR frequencies in different parts of the EPR spectrum; and **c** the local magnetic field at spin A can be manipulated by a pump pulse affecting spin B

It is assumed that one spin label in this pair, known as the "A spin", has its EPR frequency near the experimental observe frequency ω_A while the second, or "B spin", has its EPR frequency near the experimental pump frequency ω_B, Fig. 1.1b. This classification was introduced by Klauder and Anderson [4] to describe the closely-related instantaneous diffusion experiment in electron spin echo, or ESE, spectroscopy.

The two observe pulses at ω_A are separated by a time interval τ and produce an ESE signal [5, 6] at ω_A at time 2τ. Pump pulse 3 at ω_B flips the B spin at time T. This produces an "instantaneous" change in the local magnetic field from the B spin, Fig. 1.1c, shifting the EPR frequency of the A spin. This jump in the A spin EPR frequency changes the amplitude of its spin echo, i.e., the PELDOR signal. When the A and B spins do not interact, the spin echo remains unchanged. Thus, the PELDOR technique is very specific for spin-spin interactions. The amplitude $V(T)$ of the echo signal, as a function of T, traces out the distinctive characteristics of the spin-spin interactions.

Both the A and B spins respond non-linearly to the pulses, so $V(T)$ is distorted by any overlap between the pump and observe pulses or even their ringing. This means that a small, but important, piece of $V(T)$ is not measurable by the 3pPELDOR sequence. Immediately following $T = 0$, there is a short dead time, $t_{deadtime}$, where the signal is distorted and unusable.

1.1.2 Four-Pulse PELDOR

A version of PELDOR with an additional pulse applied at ω_A, Fig. 1.2, was later developed to avoid some of the dead time problems [7]. This version is known as 4pPELDOR. The first two observe pulses form a spin echo at time 2τ and the third observe pulse, at $2\tau + \tau_1$, refocuses that echo, recreating it at time $2\tau + 2\tau_1$. The pump pulse at ω_B is applied between observe pulses 2 and 3, Fig. 1.2. The amplitude of the refocused echo, $V(T)$, can be measured in the interval from $T = -\tau$ to $T = \tau_1$ where $T = 0$ corresponds to the pump pulse coincident with the ESE. Information about the spin-spin interaction can be extracted from the $V(T)$ time trace, just as for 3pPELDOR.

In the ideal case, $V(T)$ is symmetric around $T = 0$, i.e., $V(+T) = V(-T)$, and has the same shape as the 3pPELDOR $V(T)$. No observe pulse is present at $T = 0$ in 4pPELDOR, so $V(0)$ can be measured. Thus, the 4pPELDOR sequence has zero dead time, although there is a small uncertainty from the pump pulse width t_p. A similar pulse sequence is used in NMR spectroscopy to study dipole interactions between nuclei [8, 9].

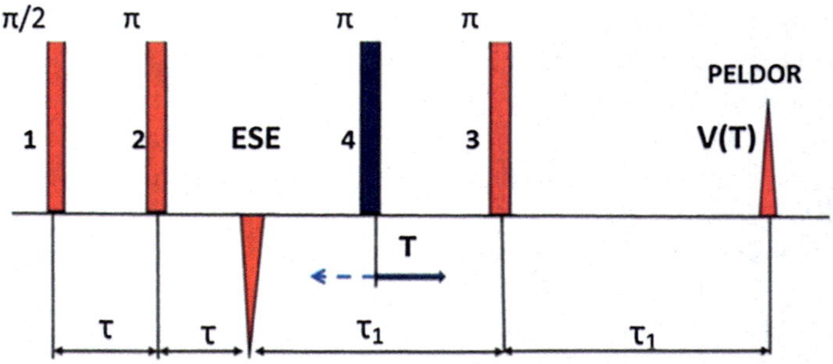

Fig. 1.2 The 4pPELDOR technique: the observe pulses (red) at ω_a at times 0, τ, and $2\tau + \tau_1$ produce a refocused ESE from *A* spins at time $2\tau + 2\tau_1$; a pump pulse at ω_b (blue) is applied at $2\tau + T$; the optimal turning angles are shown before and above each pulse

1.1.3 The Vector Model

The vector model description of the spin echo is widely known [5, 6]. PELDOR can be described in similar terms. At thermal equilibrium, the magnetic moments of the electron spins are directed mainly along the applied external magnetic field $\vec{B}_0$, Fig. 1.3a. The equilibrium magnetization, $\vec{M}_0$, depends on the static magnetic susceptibility as $\vec{M}_0 = \chi_0 \vec{B}_0$. The motion of $\vec{M}_0$ is described in an *x*, *y*, *z* axis system or coordinate frame rotating at $\omega_A (= \gamma_e B_0)$ around the *z* axis. An observe pulse

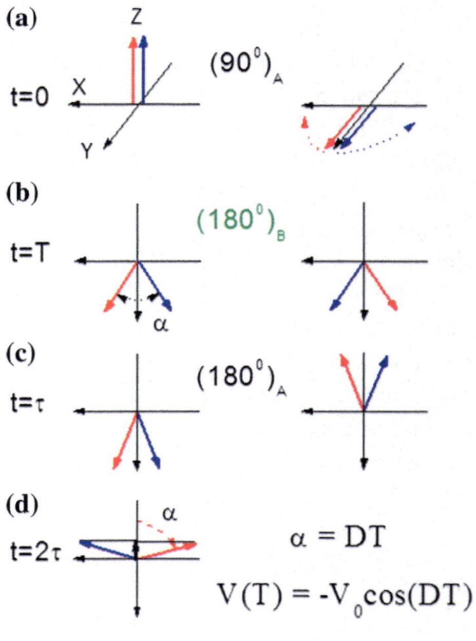

Fig. 1.3 Vector model illustrating the motion of the magnetization vector in the rotating frame in 3pPELDOR: **a** Action of the first observe pulse; **b** precession and action of the pump pulse, viewed from above the *x*, *y* plane; **c** action of the second observe pulse; and **d** formation of the PELDOR signal $V(T)$; see text for details

with its magnetic field B_1 directed along the x-axis, duration t_p, and frequency ω_a, rotates the $\vec{M}_0$ of the A spins from $\vec{B}_0$, along the z-axis, through a turning angle

$$\theta_i = \gamma_e B_1 t_{p,i} \tag{1.3}$$

toward the y-axis.

To keep discussion simple, we consider the 3pPELDOR experiment in Fig. 1.1 with observe pulses 1 and 2 having turning angles of $\pi/2 = 90°$ and $\pi = 180°$, respectively, and a pump pulse at ω_B with a turning angle π for B spins. Each A spin interacts with a single B spin, Fig. 1.1c. Each A spin falls into one of two groups or subensembles based on the local field from its B spin partner. The equilibrium magnetization of each subensemble is denoted by red or blue arrows in Fig. 1.3a, depending on whether spin interactions shift the magnetization to a higher EPR frequency (blue) or lower (red). We consider only the set of A spins that would be exactly in resonance with ω_A, were it not for the interaction D with their individual B spin. Additional broadening of the EPR absorption line, e.g. by hyperfine interactions or g-anisotropy, is readily taken into account, but makes the description considerably more cumbersome without changing the final result.

The first observe pulse rotates the magnetizations of both subensembles by $\pi/2$ around the x-axis of the rotating frame. The subensemble magnetization vectors then rotate in the x, y-plane, but in opposite directions at frequencies $\pm D/2$, Fig. 1.3a, because the resonance frequency of the A spins in each subensemble is shifted by the interaction with their B spin partners. The blue magnetization arrow of one subensemble of A spins moves counter-clockwise or faster in the rotating frame while the red magnetization arrow from the other subensemble moves clockwise or slower.

At the time of the pump pulse, the two magnetizations have rotated so that there is an angle $\alpha = DT$ between them, Fig. 1.3b. The pump pulse has no immediate effect on the A spins because the pump pulse at ω_B is resonant with, and directly affects, only B spins. However, the magnetization vectors "change their color", i.e., the A spin EPR frequency shifts by $\pm D$ because the interaction of each A spin with its partner B changes sign, Fig. 1.3b. This can also be described as an "instantaneous" exchange of magnetization between the subensembles. Regardless of how it is described, the subensemble magnetizations continue to rotate, but according to their new "color".

Just before the second observe pulse at time $\tau - 0$, the total phase of the subensemble magnetizations will be $\varphi(\tau - 0) = \pm D(\tau/2 - T)$. The second observe pulse reverses the phase of the A spins that will form the PELDOR signal. So, immediately after the pulse at $\tau + 0$, the total phase is $\varphi(\tau + 0) = \pi - \varphi(\tau - 0)$, Fig. 1.3c. Rotation continues during τ after the final pulse and adds an additional phase of $\pm D\tau/2$. At the time of the spin echo at 2τ, the phase of each subensemble is $\varphi(2\tau) = \varphi(\tau + 0) \pm D\tau/2 = \pi \pm DT = \pi \pm \alpha$, Fig. 1.3d.

In the absence of the pump pulse, the subensembles do not change "color" at time T and have a total phase shift of $\varphi(2\tau) = \pi$ and their magnetizations are focused along the $-y$ axis of the rotating frame as a conventional spin echo. The pump pulse, combined with a spin-spin interaction, causes the "refocusing" of magnetization in the conventional spin echo measurement to be inexact. The echo signal, because of the pump pulse, is modulated by the frequency of the spin-spin coupling in the spin pair:

$$V(T) = -V_0\cos(DT) \tag{1.4}$$

where $-V_0$ is the ESE signal amplitude in the absence of the pump pulse.

All the A spin magnetization at $T = 0$, immediately after the first observe pulse, is parallel to the $+y$-axis, Fig. 1.3a right-hand side. All the A spin magnetization is also parallel to the $+y$-axis at the center of the ESE in 4pPELDOR which is the point taken as $T = 0$.

Generally, only a fraction of the B spins are affected by the pump pulse. Then, the measured signal is a weighted sum of the PELDOR signal of Eq. 1.4 and the conventional spin echo signal V_0. The combined signal is often written as

$$V(T) = V_0(1 - p_B(1 - \cos(DT))) \tag{1.5}$$

with p_B the probability of a B spin being flipped by the pump pulse. Equations 1.4–1.5 explicitly show only the dependence of signal amplitude on time T. The V_0 is an implicit function of the time τ and other parameters of the observe mw pulses that are held constant during a PELDOR measurement. Similar results were obtained in the context of NMR double nuclear-nuclear resonance [10].

This very simple description shows why the PELDOR signal becomes modulated by interactions between electron spins. However, consideration of several details is required to accurately extract the interactions and distances between spins from a set of experimental PELDOR measurements.

1.2 PELDOR Theoretical Background

In this section we consider the theoretical background necessary to analyze the PELDOR time trace for various types of spatial organization of the spin system. Several basic assumptions are made in introductions to PELDOR to simplify the discussion [11]. Unfortunately, violation of these assumptions can produce artifacts and we will note some important cases. The "conventional" set of assumptions is:

1. The frequency difference between the observe and pump mw pulses $\Delta = |\omega_A - \omega_B|$ is large compared with the pulse amplitudes ω_{1A} and ω_{1B} in frequency units:

$$\Delta \gg \omega_{1A}, \omega_{1B}. \tag{1.6}$$

2. The excitation band of each pulse does not include both A and B spins. If the A and B spins have distinct, non-overlapping EPR spectra centered at ω_A and ω_B, respectively, this restricts the width of each mw pulse $t_{p,i}$:

$$t_{p,i} \gg 1/\Delta. \tag{1.7}$$

3. The splitting D in the EPR spectrum from dipole-dipole and/or exchange interactions is much smaller than the strength ω_1 of the mw fields:

$$|D| \ll \omega_{1A}, \omega_{1B}. \tag{1.8}$$

4. The undistorted ESE signal is measured, with no contamination from free induction decay signals, other echoes, ringing from the mw pulses, or baseline offsets. This requires careful phase cycling of the mw pulses or that the time τ is much longer than the characteristic decay time of the free induction signal, T_2^*,

$$\tau \gg \left(T_2^* \approx \max\left(t_{p,i}, \frac{1}{\langle\Delta\omega\rangle}\right)\right) \tag{1.9}$$

where $\langle\Delta\omega\rangle$ is the width of the inhomogeneous EPR line.

5. Paramagnetic relaxation that flips a B spin and changes its m_S value is negligible on the timescale of 2τ. If relaxation causes several B spin flips during formation of the spin echo signal, then one more B spin flip produced by the pump pulse becomes insignificant. Many, but not all, types of relaxation can be slowed by decreasing the sample temperature.

Let us consider typical values for the parameters in Eqs. 1.6–1.9 based on the original PELDOR spectrometer [1, 3, 12]: $|\omega_A - \omega_B| \approx 6 \times 10^8 \mathrm{s}^{-1}$ or $\Delta\upsilon = 100\,\mathrm{MHz}$, $t_{p,i} = 30$ or 60 ns, $\omega_1 \approx 3{-}8 \times 10^7\,\mathrm{s}^{-1}$, and $\tau = 0.5{-}2\,\mu\mathrm{s}$. For nitroxyl spin labels, $\Delta\omega \approx 10^9 \mathrm{s}^{-1}$. Thus, most conditions in Eqs. 1.6–1.9 are satisfied if the spin-spin interactions do not exceed $\sim 10^7\,\mathrm{s}^{-1}$.

Unfortunately, the second condition is typically violated in PELDOR measurements using free radicals, including nitroxyls. A significant fraction of pairs have similar EPR frequencies, so that both spins in the pair are excited by the observe pulses. In other words, some pairs contain two A spins and no B spins. This situation does not introduce new frequencies in the PELDOR time trace, but it does alter the signal amplitude and affects conclusions based on that amplitude.

When Eqs. 1.6–1.9 hold, two types of theoretical approaches are possible. The more common approach assumes that the observe and pump pulse widths are small compared to τ, i.e., $t_{p,i} \ll \tau$. This limit is valid for spins more than 1.5–2.0 nm apart. The second approach assumes that $t_{p,i} \sim \tau$, and is valid for spins 1.2–2.0 nm apart. In either case, the conventional simplifications from Eqs. 1.6–1.9 apply.

1.2.1 Oriented Pairs with Fixed Distance

We first consider the simplest case where Eq. 1.4 is valid and the PELDOR time trace has the form of Eq. 1.5 [1, 12]. Then we proceed to increasingly more complicated situations. The analysis of PELDOR data is simplest for completely oriented systems, e.g., pairs of paramagnetic centers in a single crystal or in an oriented, highly structured polymer where every spin pair is the same distance apart and makes the same angle with the applied magnetic field $\vec{B}_0$ so that every spin pair has the same value of D. In this special case, the PELDOR time trace $V(T)$ is proportional to $\cos(DT)$, possibly with some non-zero offset. The signal amplitude is modulated by the spin-spin interaction frequency D in Eq. 1.1. The PELDOR time trace for more complex systems can be derived from this simplest case by appropriate averaging over the distribution of orientations and distances.

1.2.2 Randomly Oriented Pairs with Fixed Distance

If the spins in each interacting pair are at the same distance, i.e., constant r_{AB}, but the orientation of pairs with respect to $\vec{B}_0$ is random, the PELDOR time trace is just the average of Eq. 1.5 over the angle θ. The result is [12–14]

$$V(T) = V_0(1 - p_B(1 - \Phi(T))) \tag{1.10}$$

$$\Phi(T) = \frac{\left(c(v)^2 + s(v)^2\right)^{1/2}}{v} \cos\left((\omega_D + J)T - \operatorname{atan}\left(\frac{s(v)}{c(v)}\right)\right) \tag{1.11}$$

where c and s are the Fresnel sine and cosine integrals

$$\begin{aligned} c(v) &= \int_0^v \cos\left(\frac{\pi}{2}x^2\right)dx, \\ s(v) &= \int_0^v \sin\left(\frac{\pi}{2}x^2\right)dx, \\ v &= \left(\frac{6\omega_D T}{\pi}\right)^{1/2}. \end{aligned} \tag{1.12}$$

The PELDOR time trace in Eq. 1.10 has oscillations, Fig. 1.4, due to the dipolar and exchange interactions, even in randomly-oriented systems. When $J = 0$, the time trace is just the Fourier transform of the classic Pake dipolar doublet, Fig. 1.5, with prominent oscillations at ω_D coming from the two sharp peaks of the Pake doublet.

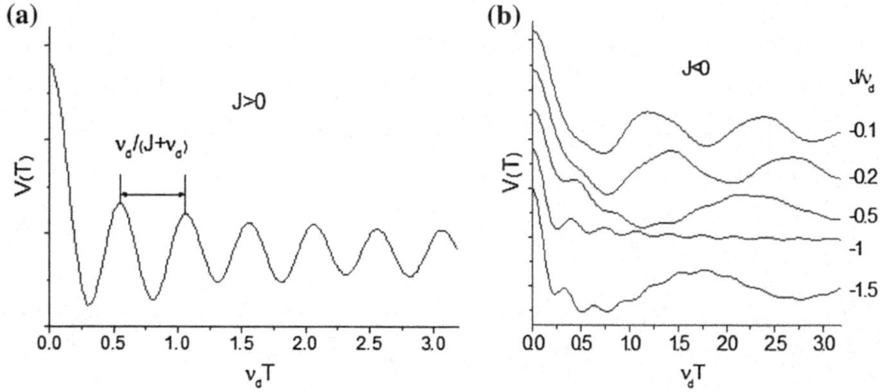

Fig. 1.4 Oscillations of the PELDOR signal resulting from dipole-dipole and exchange interactions for randomly oriented pairs of spins separated by a constant distance: **a** $J > 0$; **b** $J < 0$ with the indicated $J/\nu_D = 2\pi J/\omega_D$ ratios

Fig. 1.5 The Pake dipolar doublet for $J = 0$

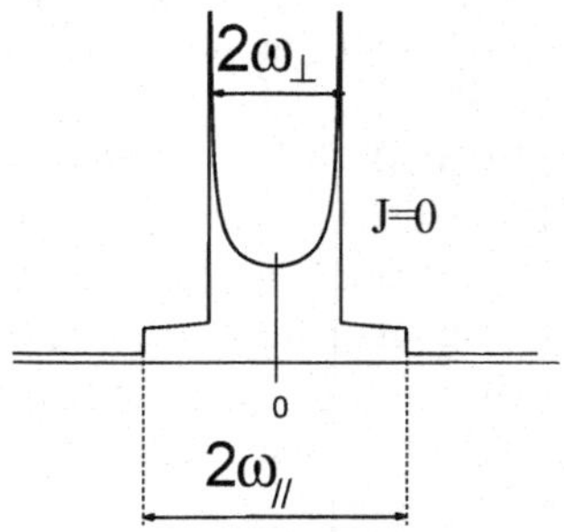

For small, positive exchange interactions, $J \geq 0$, the two prominent features of the Pake doublet at the frequencies $\pm\omega_\perp = \pm\omega_D$, Fig. 1.5, move apart to $\pm(\omega_D + J)$ while the outer edges of the Pake doublet at $\pm\omega_\parallel = \mp 2\omega_D$ move toward each other to $\mp(2\omega_D - J)$. The resulting spectrum retains the two prominent features of the Pake doublet. The Fourier transform of the spectrum is the PELDOR time trace, which retains prominent oscillations at $\omega_D + J$, Fig. 1.4a.

In contrast, the PELDOR time trace $V(T)$ becomes significantly distorted for $J < 0$. The sharp features of the Pake doublet move toward each other and start to merge into one broader feature at zero frequency, while the outer edges of the Pake doublet move apart to $\mp(2\omega_D + |J|)$. The PELDOR time trace now shows more complicated oscillations, Fig. 1.4b. For instance, with $J = -\omega_D$, oscillations are observed at $\sim 3\omega_D$. But generally, the PELDOR time trace has a number of harmonics when $J < 0$ with frequencies depending on the J/ω_D ratio, Fig. 1.4b.

The PELDOR time trace $V(T)$ can be analyzed on the basis of its visible modulation frequency only for radical pairs having a fixed distance and $J \geq 0$. Even then, a non-zero J introduces significant errors in estimating the separation distance from the frequency. The multiple frequencies seen in the PELDOR time trace with

$J < 0$ can encourage multiple interpretations. Generally, Fourier transformation of the PELDOR time trace, to convert $V(T)$ into a frequency spectrum with a Pake-like dipolar doublet, makes it possible to extract the magnitudes of ω_D and J, but also the sign of J [14]. The splittings between the features in the doublet correspond to

$$\begin{aligned} \omega_{\parallel} &= |2\omega_D - J|, \\ \omega_{\perp} &= |\omega_D + J|. \end{aligned} \tag{1.13}$$

In most cases, Fourier analysis is quite easy because the PELDOR technique restricts distance measurements to $r > 1.5\,\text{nm}$, see Sect. 2.6.3, where, with rare exception, J is negligible [15] so that $\omega_{\parallel} = 2\omega_D$ and $\omega_{\perp} = \omega_D$ follow immediately from the dipolar spectrum, Fig. 1.5. However, J can be non-zero for spins connected by a highly-conjugated linker.

The amplitude of the PELDOR oscillations decays in polyoriented systems as $\omega_D T$ increases, Eqs. 1.10–1.11. This is a simple consequence of the mutual interference of frequencies in the Pake dipolar doublet. As the doublet becomes broader, the oscillations are damped more quickly. The intense peaks in the Pake dipolar doublet are divergences whose oscillations should decay quite slowly in the PELDOR time trace. However, the oscillations in experimental PELDOR data are generally damped much faster than expected for a Pake doublet.

An important reason for the enhanced damping lies in the dispersion or spread of r_{AB}. Even a small, 10% spread, $\delta r_{AB}/r_{AB} \sim 0.1$, causes strong damping of oscillations in just three cycles, Fig. 1.6. The oscillations virtually disappear for $\delta r_{AB}/r_{AB} \geq 0.25$. Thus, simple Fourier analysis of $V(T)$ is of limited use in determining r and J.

A nonparametric method for determining the average distance between spins in a pair of paramagnetic centers with a broad distribution of distances was proposed [16], based on the position of the first minimum in the derivative of the PELDOR

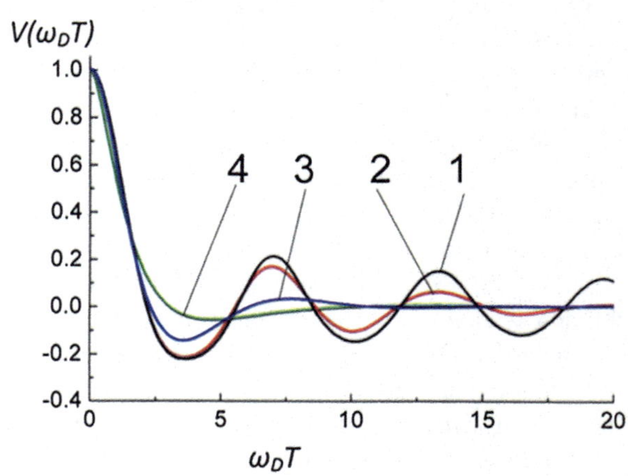

Fig. 1.6 Effect of the spread of spin-spin distances ($\delta r_{AB}/r_{AB}$) on the PELDOR time trace: $\delta r_{AB}/r_{AB}$ = (1, black) 0; (2, red) 0.1; (3, blue) 0.15; and (4, green) 0.25

time trace with respect to time T. The method requires only the initial drop in the PELDOR time trace but provides rapid evaluation of the average distance that can be used for screening or experimental setup with no assumptions about the distribution function. We return later to direct analysis of the PELDOR time trace to extract detailed information on the distance distribution function of pairs, but first we need to consider some practical aspects of the PELDOR measurement.

1.2.3 Distributions of Randomly Oriented Pairs

Most molecular systems have some flexibility or disorder, including those studied by PELDOR. It is typical to have a range, or distribution, of distances between spins in the sample. That means that the experimental PELDOR time trace is an average over the distance distribution function $F(r)$ of the time traces from each pair in the sample:

$$V(T) = V_0 \langle 1 - p_B (1 - \cos(D(r,\theta)T)) \rangle_{r,\theta}. \tag{1.14}$$

When the pairs have random orientations relative to $\vec{B}_0$, the time trace can be written in the form of a Fredholm integral equation of the first type:

$$\begin{aligned} V(T) &= V_0 \int_{r_1}^{r_2} F(r) \langle 1 - p_B (1 - \cos(D(r,\theta)T)) \rangle_{\theta} dr \\ &= V_0 \int_{r_1}^{r_2} F(r) (1 - p_B (1 - \Phi(T))) dr \end{aligned} \tag{1.15}$$

where r_1 and r_2 are limits of the range of distances between spins in a pair. The form of Eq. 1.15 makes it easy to calculate a time trace for a particular $F(r)$, but promises difficulties in attempts to recover $F(r)$ from experimental data, see Sect. 1.3.

1.2.4 Randomly Distributed Spins

Conventional PELDOR measurements require samples containing a large number of pairs of radicals. Each observed A spin interacts with all possible B spins: the partner in its own spin pair and all other spins in the sample. The spins in other pairs are generally scattered at random throughout the sample at all possible distances and angles. These randomly-distributed spins can act as B spins to affect the A spins.

The PELDOR time trace from interactions with all these potential B spins at random r and θ has been calculated using the Markov method assuming that these B spins independently affect the A spins [17, 18]. Their PELDOR time trace is

$$V(T) = V_0 \left\langle \prod_j \left(1 - p_B\left(1 - \cos\left(D_j T\right)\right)\right) \right\rangle_{r,\theta} \tag{1.16}$$

where the subscript j refers to every B spin and the angular brackets denote averaging over angles and distances. The average of Eq. 1.16 for a random spatial distribution of spins produces an exponentially-decaying PELDOR time trace

$$\begin{aligned} V(T) &= V_0 \exp\left(-\frac{8\pi^2}{9\sqrt{3}}\gamma^2 \hbar p_B C T\right) \\ &= V_0 \exp\left(-2 p_B \Delta\omega_{1/2} T\right) \end{aligned} \tag{1.17}$$

where $\Delta\omega_{1/2} = 8.2 * 10^{13} \mathrm{C\,cm^3\,s^{-1}}$ is the dipolar linewidth and C is the concentration of paramagnetic centers. This PELDOR result corresponds exactly to the expression for the ESE decay due to dipolar interactions of randomly distributed particles, known as instantaneous diffusion [4, 6].

However, spins can be distributed in inhomogeneous media, such as micelles, membranes and surfaces. Then the average of Eq. 1.16 over r and θ must take into account the actual distribution of B spins with respect to an A spin. One important class of such distributions are characterized by a fractal dimension d characterizing the spatial distribution of spins. That is, the probability of there being a B spin a distance r from an A spin is proportional to $r^{d-1}dr$, at least over the range of distances accessible to PELDOR. For such a fractal distribution, the average of Eq. 1.16 gives

$$V(T) = V_0 \exp\left(-\alpha p_B T^{d/3}\right) \tag{1.18}$$

where the dimensionality d of the spatial distribution of spins appears as a power in the exponential function in Eq. 1.18. For instance, a simple exponential decay with dimensionality of $d = 3$ occurs for a random distribution of spins in a 3-dimensional volume; $d = 2$, a distribution on a plane; and $d = 1$, a linear distribution [18, 19]. The dimensionality for complex spatial distributions of spins can be calculated using Monte-Carlo methods [20]. For a spin system with a fractal dimension d, the concentration of spins C in Eq. 1.17 is replaced by $C(r)$ which depends on distance as [21, 22]

$$C(r) = k_d r^{d-3} \tag{1.19}$$

with k_d absolutely independent of r. The effective PELDOR relaxation rate α in Eq. 1.18 has the form

$$\alpha = p_B \frac{8\pi}{3 2^{d/3}} k_d \left(\gamma^2 \hbar\right)^{d/3} \mathrm{I}_d \tag{1.20}$$

$$\mathrm{I}_d = \int_0^\infty \frac{\sin(t)^2}{t^{(d-3)/3}} dt \int_0^1 \left(1 - 3x^2\right)^{d/3} dx \tag{1.21}$$

$$k_d = \mathrm{C}_d \frac{d}{4\pi} \frac{\Gamma\left(\frac{1}{2}\right)^d}{\Gamma\left(1 + \frac{d}{2}\right)} \tag{1.22}$$

where C_d is the d-dimensional spin concentration of the radical distribution function; and $\Gamma(x)$ is the Euler gamma function. For a homogeneous, random distribution of spins in a 3-D volume, C_d is the standard radical concentration C.

The dimensionality of the spin distribution also affects the echo decay in ESE studies, but the effect is typically masked by other relaxation mechanisms [6]. The dimensionality of the spin distribution is quite prominent in PELDOR time traces and must be considered in the analysis of PELDOR data. In fact PELDOR measurements provide a unique opportunity to study the dimensionality of the distribution of labeled molecules, e.g., to determine if the labeled molecules are confined to a volume, to a surface or membrane, or to fibrils or extended chains.

1.2.5 *Intra- and Inter-molecular Dipolar Interactions*

The dipolar interactions in complex systems, e.g., containing spins in pairs, clusters or aggregates, are often divided into two classes. There are interactions among radicals within the same cluster, often called intramolecular or intra-cluster interactions, and those between radicals of different clusters, the intermolecular interactions. If the interactions are independent of each other, the PELDOR time trace is the product of the time traces for the two types of contributions

$$V(T) = V_{INTRA}(T) V_{INTER}(T). \tag{1.23}$$

This makes it possible to focus on the dipolar interactions of interest in complex systems. For instance, in frozen solutions of biradicals, Eqs. 1.10–1.12 describe the intra-pair dipolar contribution $V_{INTRA}(T)$, while Eq. 1.17 often describes the inter-pair interaction, $V_{INTER}(T)$. The two contributions can be extracted from PELDOR time traces measured at different concentrations of radical pairs [23, 24]. However, $V_{INTER}(T)$ may be rather different even from Eq. 1.18, and the convenient approximation of $V_{INTER}(T)$ by a simple exponential, as is very often done, can be quite dangerous. One should determine with certainty, rather than hope, that there are no spatial inhomogeneities, fractal effects or other complicating factors. In particular, ionic or electrostatic effects alter the mutual distribution of clusters,

Sect. 5.2 [25], and excluded volume effects prevent the close approach of macromolecules Sect. 5.4 [2, 21], and produce $V_{INTER}(T)$ rather different from the simple exponential expressions Eqs. 1.17–1.18. The $V_{INTRA}(T)$ extracted from $V(T)$ using an incorrect $V_{INTER}(T)$ causes errors and false maxima in the reconstructed pair distribution function.

1.2.6 Clusters of N Spins

It is not infrequent that distances between spins within a cluster are much shorter than the distances between clusters as in, e.g., radical pairs formed by photolysis or radiolysis of solids, multiple spin labels in a protein complex, or in synthetic biradicals. In some cases, the clusters have a highly-defined structure with the same number of spins in every cluster and specific distances between spins. In other cases, clusters have a distribution of numbers of spins and distances between spins. In both extremes, the analysis of PELDOR time traces can help characterize the geometry and size of such clusters [12, 13, 26].

For simplicity, we will consider a set of clusters, each with an identical number of spins N. More disperse systems can be considered a mixture of well-defined clusters. We assume that the position and orientation of a spin is unrelated to its distance from the other spins within that cluster. If the observe pulses excite only one spin per cluster, the PELDOR time trace due to intra-cluster dipolar interactions is

$$V_{INTRA}(T) = V_0\Big(1 - p_B\langle 1 - \cos(DT)\rangle_{r,\theta}\Big)^{N-1} \tag{1.24}$$

The average over distance and angle between spins within each cluster leads to rapid decay of $V_{INTRA}(T)$ at short times

$$T \leq T^* \approx \frac{r_{eff}^3}{\gamma^2\hbar} \tag{1.25}$$

where r_{eff} is an effective separation between spins within a cluster. When $p_B \ll 1/N$, only rarely is more than one B spin in a cluster excited by the pump pulse. In this limiting case, the Fourier transform of Eq. 1.24 is the powder average spectrum of dipolar splittings between A and B spins, and enables extraction of the distance distribution function f_N for the cluster.

The normalized $V_{INTRA}(T)/V_0$ in Eq. 1.24 approaches its limiting value of V_p for $T \geq T^*$, Fig. 1.7a, where, to a good approximation, $\langle\cos(DT)\rangle \to 0$, and the limiting V_p is [12, 26]

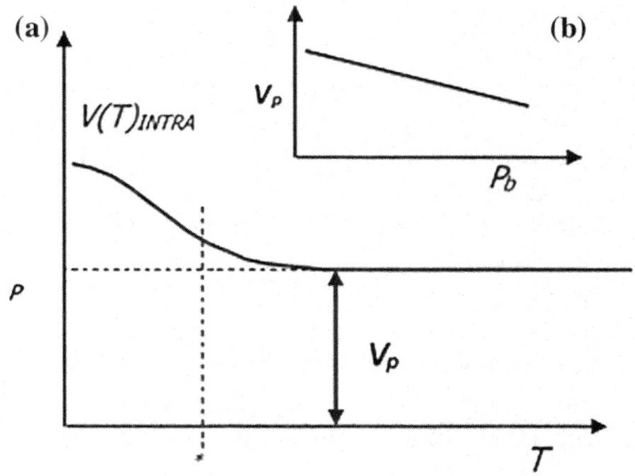

Fig. 1.7 **a** PELDOR time trace for N-spin clusters with no correlation between spin-spin distances within each cluster; **b** the dependence of the limiting normalized V_p on the pump pulse excitation probability p_B, showing a slope of $-(N-1)$

$$V_p = (1 - p_B)^{N-1} \cong 1 - (N-1)p_B. \tag{1.26}$$

The approximate value on the right-hand side is valid for $Np_B < 1$; so that the number of spins N in the cluster is available from the experimentally determined V_p and p_B, Fig. 1.7b.

The observe pulses present a complication in determining N when there is more than one potential A spin in a cluster. Dipolar modulation of the ESE occurs if more than one A spin in a cluster is excited by observe pulse 2 for 3pPELDOR or pulses 2 or 3 for 4pPELDOR [27–30]. Dipolar modulation of the ESE alters the intensity of the PELDOR signal and the value of V_p. Keeping $p_A \ll 1$ for observe pulses 2 and 3 eliminates this problem but can greatly reduce sensitivity. But even when p_A is significant, N can be determined from 3pPELDOR measurements using Eq. 1.37.

1.2.7 Statistical Distribution of Cluster Sizes

If all clusters do not have the same number of spins N, then Eq. 1.24 must be averaged over the distribution of N. The resulting average of products of averages of PELDOR modulations thoroughly entangles the distributions of distances and cluster sizes. It is generally not possible to extract either distribution function. However, it is possible to characterize the distributions statistically in terms of their moments, as described later in this section.

A few common situations with several spins per cluster include: spins absorbed on small particles; defects in small particles; radiolysis products in solids; and redox centers in large protein complexes. Spins in such clusters often have similar or identical EPR spectra, so it is difficult to make a clear division into A and B spins. Generally, several spins in the same cluster are potential A spins; and the pump pulse may excite more than one B spin. As a result, the "conventional" assumptions for PELDOR analysis summarized in Eqs. 1.6–1.7 are typically violated.

When p_B is not small, i.e., $p_B \gtrsim 1/N$, the pump pulse excites more than one B spin in a cluster. The resulting $V_{INTRA}(T)$ from Eq. 1.24 contains products of cosines which can be expanded as cosines of sums and differences of dipolar splittings from the excited B spins, entangling the distance information. The Fourier transform of the PELDOR time trace for large p_B approaches the dipolar lineshape, which typically has a Gaussian form if several spins are involved. A single parameter, i.e., the linewidth or second moment, completely describes that Gaussian form and that single parameter is insufficient to determine N and/or f_N.

For small $p_B \ll 1/N$, the shape of the PELDOR time trace in Eq. 1.24 can provide some additional information, as seen in Sect. 1.2.6, which survives the averaging over cluster sizes. The Fourier transform of the averaged time trace in the small p_B limit is the average of dipolar spectra of all spins in all cluster sizes, giving distance information in the form of the first few moments of the distribution of dipolar splittings between pairs of spins. When combined with the second moment of the sum of the dipolar splittings of all spins in all cluster sizes, which is obtained from measurements at large p_B; the average number of spins per cluster, N can be extracted. Information about N also appears in the limiting value of the PELDOR time trace, V_p, Eq. 1.26; but the observe pulses complicate matters, as noted in Sect. 1.2.6.

The use of weak pulses to avoid exciting multiple A spins achieved some success in 3pPELDOR studies of distributions of moderately sized clusters [27, 28]. The probability that the second observe pulse in 3pPELDOR excites only one spin in a cluster of n spins is $f_N(n)np_2z^{n-1}$ where $z = 1 - p_2$ and p_2 is the probability that an A spin is flipped by the second observe pulse. For small p_2, large clusters contribute more to the PELDOR signal than do small clusters, particularly because instantaneous diffusion effects are minimized. The PELDOR time trace in Eq. 1.24 can be summed over cluster size to give a limiting value at long T of [27, 28]

$$V_p = 1 - \overline{N}p_B \tag{1.27}$$

in the limit of small p_B. The average cluster size, i.e., the first moment of the distribution of N, comes from

$$\bar{N}(z) = \frac{\sum_{n=1}^{N_{\max}} f_N n^2 p_2 z^{n-1}}{\sum_{n=1}^{N_{\max}} f_N n p_2 z^{n-1}} \tag{1.28}$$

where N_{max} is the maximal number of spins in a cluster. Thus, it is possible to experimentally determine $\overline{N}$ as the limiting value of Eq. 1.28 as z approaches 1 and to estimate the first moment K_1 and the second moment K_2 of the distribution function f_N as:

$$K_1 = \sum_{n=1}^{N_{\max}} n f_N = \overline{N}(1) \tag{1.29}$$

and

$$K_2 = \sum_{n=1}^{N_{max}} (n - K_1)^2 f_N = \frac{d\overline{N}(z)}{dz}. \tag{1.30}$$

This approach does yield moments of the distance distribution function f_N [27, 28], but seems limited to 3pPELDOR of systems with intense EPR signals because of the requirement to extrapolate to the limit $p_2, p_B \rightarrow 0$ where signals become quite weak and noisy. In 4pPELDOR, the third observe pulse must also be weak, further diminishing the signal.

1.3 Reconstructing the Distance Distribution Function

Methods to determine the distance distribution function $F(r)$ from PELDOR data are still being actively developed [2, 16, 31–37] and several approaches have had some success. The major focus in these efforts are radical pairs, i.e., clusters consisting of two spins, because of the many PELDOR studies of spin-labelled biomolecules and because PELDOR time traces of larger clusters converge to that of pairs, as seen in Sects. 1.2.6 and 1.2.7. Each approach faces similar challenges and each has some limitations. In later chapters, a number of analysis methods are mentioned during discussions of PELDOR studies. Here, we outline some of the challenges and issues involved in obtaining a reliable distance distribution from PELDOR data.

1.3.1 The Inverse Problem

Let us start by considering idealized PELDOR data where pairs of radicals separated by r produce a Pake pattern and a PELDOR time trace given by Eq. 1.11 with $J = 0$. The time trace for pairs of such radicals with a distribution of distances is just the sum or integral of Eq. 1.11 weighted by the distance distribution $F(r)$. It is easy to start with $F(r)$ and then calculate the time trace, but the inverse, starting from a time trace and calculating the $F(r)$, turns out to be much more difficult.

The inverse calculation faces a fundamental difficulty. Shapes of the PELDOR time traces, Eq. 1.11, for pairs separated by similar distances, e.g., r and $r + \delta r$ for tiny δr, are virtually identical and the differences between them are much smaller than the experimental noise. Consequently, the noise or error in the experimental

PELDOR time traces overwhelms the subtle differences needed to correctly determine the relative values of $F(r)$ and $F(r+\delta r)$. In fact, many distance distributions, with substantial differences between each other, will reproduce the time trace to within the experimental noise level. Many of these distance distributions are physically unreasonable, with negative values at some distances or unrealistic, rapid variations with distance. Simple least-squares fitting methods often produce physically unreasonable $F(r)$, a phenomenon known as "over-fitting" by some PELDOR spectroscopists. A simple indication of overfitting is that two identical measurements, differing only in noise, yield quite different distance distributions.

The determination of $F(r)$ from the PELDOR time trace is characterized in mathematical terms as an ill-posed inverse problem. One approach is to force $F(r)$ to have a 'suitable' distribution, e.g., the sum of a few Gaussian distribution curves. Another approach uses regularization techniques, such as those developed by Tikhonov [38], to suppress unreasonable behavior, e.g., negative values of $F(r)$. Every well-characterized approach is known to have difficulties with certain types of distributions, e.g., Tikhonov regularization does not perform well for distributions having a mixture of narrow and broad peaks. In one sense, these approaches do not solve the fundamental difficulty; they simply avoid distribution functions that are obviously unsuitable. One needs to keep in mind that the instability of the inverse problem is usually retained: slight changes in the PELDOR time trace, e.g., from a different sampling of random noise, can produce large changes in the calculated $F(r)$.

1.3.2 Steps in the Reconstruction of $F(r)$

The analysis of PELDOR measurements generally focuses on determining the distance distribution function $F(r)$ for spins in a cluster. Some studies have examined the distribution of spins in solution, or attached to particles, surfaces or polymers; these are described in Chap. 5. We will describe one approach to analysis of data for a cluster of N spins in the limit of small p_B where the PELDOR time trace converges to that of a pair of spins, as described in Sect. 1.2.6.

We will assume that the inter-cluster contribution $V_{INTER}(T)$ to the time trace has already been removed from the data, see Sect. 1.2.5. However, $V_{INTER}(T)$ can deviate from Eq. 1.18 at small T from excluded volume and electrostatic effects, particularly with biomacromolecules, Sect. 5.2. This region of $V_{INTRA}(T)$ is quite important for reconstructing $F(r)$ and there are strong recommendations to determine $V_{INTER}(T)$ and $F(r)$ simultaneously from $V(T)$ [2].

The distance distribution between spins is defined as $F(r) = dn(r)/dr$, where $dn(r)$ is the proportion of pairs of spins within the cluster having separations between r and $r+dr$. The $F(r)$ is typically determined by comparing the experimental $V_{INTRA}(T)$ with calculated $V(T)$ for clusters containing N spins.

The experimental PELDOR time trace $V_{INTRA}(T)$ is conveniently analyzed from the normalized PELDOR time trace

$$V_n(T) = \frac{V_{INTRA}(T) - V_p}{V_0 - V_p} \tag{1.31}$$

where V_0 is $V_{INTRA}(0)$ in the absence of clusters, often regarded as the signal in the absence of the pump pulse, and V_p is the limiting value of the time trace at large T. However, V_0 cannot be measured experimentally by simply turning off the pump pulse. That procedure introduces two experimental artifacts. First, the pump pulse could be exciting the observed A spin, decreasing its signal. Second, the pump pulse produces a dynamic phase shift of the PELDOR signal and splits it into three echoes, having different phases and shifted in time from each other by approximately the pump pulse width [39]. The signal measured by turning off the pump pulse will differ from V_0 by a shift in phase, by interference between the overlapping echoes, and by the direct excitation of the A spins by the pump pulse. Rather, V_0 must be determined by fitting the PELDOR signal and extrapolation to $T = 0$.

Ideally, no more than one pair of spins in each cluster contributes to the PELDOR signal and the PELDOR time trace matches that of randomly oriented pairs of spins, Eqs. 1.10–1.12. This assumption is valid for clusters with $N = 2$ or in larger clusters with small p_B and p_A, so that the probability of exciting more than two spins is negligible. Unfortunately, these conditions also make the signal for large clusters quite weak.

For a continuous distribution of distances between spins in such clusters, the calculated signal is

$$V_{CALC}(T) = V_p + (1 - V_p)p_B \int_{r_1}^{r_2} F(r)f(r,T)dr \tag{1.32}$$

where $V_p = 1 - (N - 1)p_B$ and $f(r,T) = \Phi(T)$ from Eqs. 1.11 and 1.2.

The integration limits r_1 and r_2 in Eq. 1.32 are the physically reasonable range of distances between spins in a cluster. The integral can be broken into a sum over discrete values r_k with widths dr_k. The $V(T)$ at the experimental set of pump delays T_S is

$$V_{CALC}(T_S) = V_p + (1 - V_p)p_B \sum_{k=1}^{m} F(r_k)\left(\int_{r_k}^{r_k+dr_k} f(r,T_S)dr\right)dr_k \tag{1.33}$$

where m is the number of shells between r_1 and r_2.

The best distribution function $F(r_k)$ gives a $V_{CALC}(T_S)$ that comes close enough to the experimental $V_n(T_S)$, subject to any constraints we want to place on $F(r_k)$. The constraints, sometimes characterized as prior information, can be applied through the Tikhonov regularization method, whose applicability to PELDOR data analysis was thoroughly examined [34]. Various constraints have been successfully used, including constraints on the distance distribution function, e.g., $F(r_k) \geq 0$ or minimal $|F(r_k)|$, or on its smoothness, e.g., minimizing the gradient or first- or second derivative, or on the functional form of $F(r_k)$, e.g., a sum of Gaussian peaks. The relative merits of some of these approaches were recently examined [31].

The Tikhonov regularization approach uses a parameter λ [34] to balance the relative importance of reproducing the PELDOR time trace at $\lambda \to 0$ against satisfying the imposed 'suitability' constraints at $\lambda \to \infty$. The resulting $F(r)$ depends on the regularization parameter λ. For small λ, the fitting remains an ill-posed inverse problem and $F(r)$ may vary chaotically with large, sudden changes in amplitude. For large λ, the solution lacks important details and may become independent of the experimental data. There are various competing criteria for selecting the optimum $F(r)$ and λ [31, 34].

This is a brief outline of a general approach to determine $F(r)$, which has been used by multiple authors. Programs for reconstruction of $F(r)$ from PELDOR data are widely available (for a detailed description, see [2, 40]). We believe that one aspect of this ill-posed inverse problem still needs to be remembered: small changes in experimental data can produce large changes in $F(r_k)$, even when Tikhonov regularization delivers a reasonable and pleasant-looking $F(r_k)$.

Systematic or random errors in the experiment or its analysis should be carefully considered because they can bias outcomes of the analysis. The separation of the experimental PELDOR time trace into V_{INTRA} and V_{INTER} and the proper values of p_B and p_2 are critical; but so are assumptions that $J = 0$ or that orientation selection or excluded volume effects are absent; as are offsets of the T axis, drifts in spectrometer tuning or overlap of pump and observe pulses.

Other approaches to determine $F(r)$ from PELDOR data have been effective and method development continues. For example, the Tikhonov regularization step has been avoided by using Monte Carlo calculations [36] or the Mellin integral transform [37].

1.4 Orientation Selection in PELDOR

In this chapter, we repeatedly assume that the orientation of a radical or of a pair of radicals has no effect on its interaction with the mw pulses or on the detectability of its EPR signal. This assumption means that p_A and p_B do not depend on orientation and allows the PELDOR time trace to be calculated as a simple average over the dipolar splitting D, giving a simple form to $f(r, T)$ and Eq. 1.33. This is a good assumption when isotropic interactions control the EPR spectrum, e.g., in deuterium atoms where the hyperfine and Zeeman interactions are both isotropic. Then the classification into A and B spins is controlled by the large deuterium hyperfine interaction. All A spins have identical EPR spectra and all possible orientations in space, while at the same time, all B spins have identical spectra and all possible orientations.

Yet, many radicals are anisotropic and the EPR frequency of each spin is determined by their orientation in the external magnetic field. The observe and the pump pulses in PELDOR are applied to different regions of the EPR spectrum. Consequently, the *A* spins have their EPR frequency near ω_A and have a restricted range of orientations relative to external magnetic field; and similarly, the *B* spins have a restricted range of orientations. This situation is commonly called orientation selection in PELDOR and other forms of EPR spectroscopy and can have major impacts on experimental data.

For convenience, we will discuss orientation selection in the context of nitroxyl spin labels although it can occur with any anisotropic spin. Orientation selection is an important concern in PELDOR because it can alter the shape and frequency content of the PELDOR time trace and will impact its analysis.

Anisotropic spins with different spatial orientations generally are excited to different extents by the observe pulses that form the spin echo signal and by the pump pulse. So the PELDOR signal from a pair or anisotropic spins depends on the orientation of each of the spins. If the orientations of the two spins in a pair are uncorrelated with one another, p_A and p_B correctly account for the effects of the anisotropy. On the other hand, if the orientations of the two spins are correlated, as illustrated in Fig. 1.8a, the values of some parameters involving $V(T)$, p_B and D in Eq. 1.5 are correlated, so that [33, 41]

$$V(T) = V_0\langle G(\Phi, \Phi_1)(1 - p_B(\Phi, \Phi_2)(1 - \cos(D(\theta, r)T)))\rangle_{\Phi,\Phi_1,\Phi_2}. \tag{1.34}$$

The function $\Phi(\theta, \phi)$ defines the orientation of $\vec{B}_0$ in the molecular frame, Fig. 1.8. The z-axis is selected along $\boldsymbol{r}_{12}$ and the Euler angles $\Phi_i(\alpha_i, \beta_i, \gamma_i)$, define

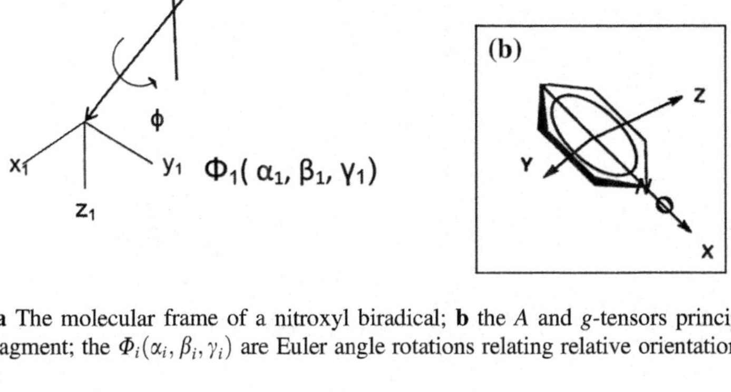

Fig. 1.8 **a** The molecular frame of a nitroxyl biradical; **b** the *A* and *g*-tensors principal axis of nitroxyl fragment; the $\Phi_i(\alpha_i, \beta_i, \gamma_i)$ are Euler angle rotations relating relative orientations, see text for details

the orientation of the hyperfine (A_{xx},A_{yy},A_{zz}) and g-tensors (g_{xx},g_{yy},g_{zz}) for each nitroxyl group, Fig. 1.8b. The function $G(\Phi, \Phi_1)$ describes the degree of spectral excitation of spin 1 which serves as the *A* spin. Fortunately, some simplifications are possible in many experimental situations.

Orientation selection is most significant when spin pairs have rigid conformations of nitroxyl labels relative to $\boldsymbol{r}_{12}$, giving a highly restricted set of Φ_i. Nitroxyl radical spins are sufficiently close to having axial symmetry, Fig. 1.9, that X- or Ku-band experiments typically need to consider just θ and only the angles β_1 and β_2 are significant parameters [41].

However, in totally-rigid biradicals or poly-radicals, the orientations of the spins are fixed relative to each other and to the vector $\boldsymbol{r}_{12}$; and the extraction of distances and numbers of coupled spins at high magnetic fields becomes quite demanding. Exactly which spin pairs contribute to the echo and to $V(T)$ depends on θ. Consequently, the Pake pattern is distorted and incomplete because it is not a complete powder average of all possible orientations. The modulation depth parameter V_p also depends on θ through p_B. Thus, methods that successfully analyze PELDOR measurements when there is no orientation selection, fail to produce correct results when orientation selection is significant. In fact, a correct distance r_{12} cannot be extracted from the observed dipolar frequency spectrum, if the relative orientations of the spins are unknown. Fortunately, the observe and pump frequencies can be varied to observe and pump different sets of orientations, Fig. 1.9, to unravel the orientational and distance information.

The correlation effects for nitroxyl radicals used in spin labeling considerably diminish at X- or Ku-band frequencies. The typical side-chain flexibility, combined with unresolved inhomogeneous broadening, result in a broad range of Φ_i resulting in a rather uniform value for *G*. The orientation selection is completely

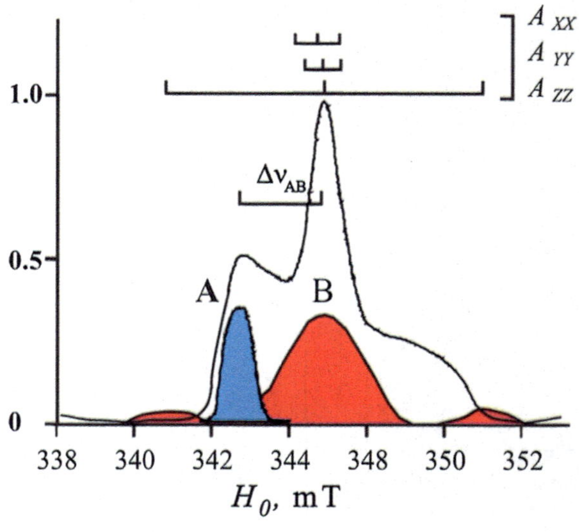

Fig. 1.9 EPR spectrum of a nitroxyl radical at X-band and typical excitation profiles of observe (blue) and pump (red) pulses: the A_{xx}, A_{yy}, A_{zz} lines indicate the ^{14}N anisotropic HFI in the spectrum; *A* and *B* refer to the observe and pump pulses

eliminated if the spins within pairs are randomly oriented, which commonly occurs for nitroxyl spin labels with sufficient conformational freedom.

Complete elimination of orientation selection simplifies PELDOR analysis, but all orientational information is lost and only the distance $\boldsymbol{r}_{12}$ between spins and its distribution function $F(r)$ can be determined. On the other hand, strong orientation selection is much more difficult to analyze correctly, but provides much more information about the spatial relationship between the two spins; up to the complete 3D organization of the radical pair. All this information can be extremely valuable and informative in totally-rigid biradicals or highly-ordered proteins, Sect. 8.2.2. Strong orientation selection is easier to achieve experimentally at higher mw frequencies and high magnetic fields (Q-, W-bands, or higher), where g-factor anisotropy begins to dominate the EPR spectrum. Then it becomes possible to use rather selective excitation of different parts of the spectrum which correspond to different orientations of the spins.

A strong indication of orientation selection is that the PELDOR time trace changes shape when different parts of the EPR spectrum are observed or pumped. If the changes are large, the full orientation and distance information may be obtainable from $V(T)$ traces covering the entire spectrum. An individual trace under orientation selection lacks some part of the full dipolar spectrum. The systematic presence and absence of this spectral information reveals the full orientation of the radical pair.

If orientation selection is weak, it may not be possible to completely determine the orientation of the pair. Yet, it is possible to recover distance information although each individual time trace lacks a part of the full PELDOR information because of orientation selection. The full dipolar spectrum can be recovered from multiple PELDOR traces.

For example, consider a PELDOR trace where a portion of the Pake dipolar doublet near $\pm\omega_{\|}$, Fig. 1.5, is lost to orientation selection effects. Its dipolar spectrum is nearly identical to the dipolar spectrum without orientation selection for a small, positive value of J and a different ω_D. That single trace could be interpreted in several ways. However, the missing portion of the dipolar spectrum does appear in other PELDOR traces measured with different orientation selection effects. That additional information about the $\pm\omega_{\|}$ region of the dipolar spectrum will resolve this problem.

One approach to obtain the full dipolar spectrum and the correct distance under orientation selection is to analyze the average of multiple PELDOR time traces, each normalized by the EPR absorption intensity at that point. Several methods for orientation selection averaging have been developed [32, 42, 43] to obtain the distance distribution function.

Several detailed measurements and analyses show the structural detail that can be achieved from PELDOR with strong orientation selection. High-frequency PELDOR measurements of a cofactor pair in photosynthetic reaction centers [44–46] are discussed in Sect. 8.2.2. Orientation selection was analyzed for high-frequency PELDOR for several other spin pairs [47–49]. At X-band, orientation selection has been used for Cu^{2+}-Cu^{2+} [50, 51], Cu^{2+}-nitroxyl [50–52] and iron-sulfur center-nitroxyl [51] pairs.

The detailed analysis and geometrical interpretation of orientation selection data are quite challenging and requires input from other structural or computational methods but the results can be well-worth the effort.

1.5 Spectral Overlap

The conventional assumptions introduced in Sect. 1.2 were extremely helpful in developing a qualitative understanding of PELDOR. That qualitative behavior persists even when those simplifying assumption are not fully met. However, in quantitative measurement of distances and cluster properties from PELDOR data, small inaccuracies in the analysis can have a significant impact on the conclusions from a study.

The first two conventional assumptions in Sect. 1.2 are commonly violated, particularly in studies where only nitroxide spin labels are used. The EPR spectra of all spins in a pair or cluster completely overlap and the classification of *A* spins and *B* spins becomes somewhat arbitrary. Fortunately, other pulse EPR methods have faced this problem and the effects are well understood. Only a few adjustments to the analysis of PELDOR measurements are necessary. The effect of spectral overlap was examined in detail for pairs of spins in 3pPELDOR [29] and 4pPELDOR [30] experiments and extension to larger clusters was considered.

Each mw pulse has four potential effects on a pair of radicals. Each pulse can excite: only the first spin, only the second spin, both spins or neither spin. There are $4^3 = 64$ possibilities with the three mw pulses in 3pPELDOR and $4^4 = 256$ in 4pPELDOR. Fortunately, most of these possibilities can be ignored. In order to contribute to the measured PELDOR signal, a spin must not be excited by the pump pulse but must be excited by all of the observe pulses; otherwise any signal it does produce occurs at a different time and will not be recorded. Also, the PELDOR signal from an *A* spin does not depend on whether its partner is excited by the first observe pulse. This conveniently reduces the number of situations that must be considered to four in 3pPELDOR and eight in 4pPELDOR.

The 3pPELDOR signal comes only from a spin excited by both observe pulses and depends on how the partner spin in the pair is excited by the second observe pulse and the pump pulse. If neither pulse excites the partner spin, any dipolar interaction is completely refocused in the spin echo, making the full contribution of 1 to the PELDOR signal, case 1 in Table 1 of [29]. If the partner spin is inverted by the second observe pulse, case 2 [29], the echo is reduced to $\cos(D\tau)$; this case corresponds to dipolar modulation of the two-pulse echo, but here τ is constant. If the partner spin is inverted only by the pump pulse, case 3 [29], the echo is modulated by $\cos(DT)$, the idealized PELDOR modulation in Eq. 1.4. Finally, if the partner spin is excited by both the pump pulse and the second observe pulse, case 4 [29], the signal is modulated by $\cos(D(\tau - T))$. The 3pPELDOR signal is simply the sum of the signals from each case weighted by the probability that the partner spin was flipped in that fashion. The results for 4pPELDOR are similar, with eight cases to be considered, and modulations involving the product of D with combinations of τ_1, τ_2 and T [30]. Fortunately, the additional terms introduced by spectral overlap do not alter the modulation frequency of the PELDOR time trace when spins occur in pairs.

Let's consider the simple 3pPELDOR measurement on pairs of spins having strongly overlapping spectra, e.g., pairs of nitroxide radicals, or pairs of trityl radicals. The EPR spectra of both radicals in a pair overlap and it is quite possible for both radicals in the pair to be excited by one or more of the observe pulses. In addition, the spectral width is limited and for some combinations of mw pulse frequency and pulse width, the observe and pump pulses might excite the same spin. The conventional assumptions in Eqs. 1.6–1.7 are violated, but the qualitative features of PELDOR are still recognizable. However, quantitative analysis to extract cluster properties requires close attention to experimental details.

The 3pPELDOR signal from such a pair of identical spins is given by Eq. (84) of [29], which can be written as

$$\begin{aligned} V(T) = V_0\Big(1 &- (p_A - p_{AB})\langle 1 - \cos(D\tau)\rangle_{r,\theta} \\ &- (p_B - p_{AB})\langle 1 - \cos(DT)\rangle_{r,\theta} \\ &- p_{AB}\langle 1 - \cos(D(\tau - T))\rangle_{r,\theta}\Big) \end{aligned} \tag{1.35}$$

when there is no orientation selection. Note that here p_A is the probability that the partner spin is flipped by the second observe pulse at ω_A, while p_{AB} is the probability that the partner spin is flipped by that second observe pulse at ω_A and by the pump pulse at ω_B. Typically $p_{AB} \ll p_A p_B$ [29], so that Eq. 1.35 converges to $V_{INTRA}(T)$, Eq. 1.24 for a pair of spins, $N = 2$, as $p_A, p_{AB} \to 0$. When p_{AB} is not negligible, dipolar modulation from the last term in Eq. 1.35, reappears near the end of the PELDOR time trace, growing in intensity as $T \to \tau$.

The partner spin is flipped by the second observe pulse when $p_A > 0$, which decreases $V(0)$ and increases $V_p = V(\infty)/V(0)$, Fig. 1.7a and Eq. 1.26. Such intensity variations have the potential to introduce errors when estimating the number of spins per cluster using Eq. 1.26. However, when $p_{AB} \to 0$, which happens when the pump and observe pulses have non-overlapping excitation bands, then the limiting value in Eq. 1.26 becomes

$$V_p = \frac{V_0\Big(1 - p_A\langle 1 - \cos(D\tau)\rangle_{r,\theta} - p_B\langle 1 - \cos(DT)\rangle_{r,\theta}\Big)}{V_0\Big(1 - p_A\langle 1 - \cos(D\tau)\rangle_{r,\theta}\Big)}. \tag{1.36}$$

At large DT, the $\langle\cos(DT)\rangle_{r,\theta}$ converges to zero as it does in Eq. 1.26. But $\tau > T$, so $\langle\cos(D\tau)\rangle_{r,\theta}$ also has converged to zero and Eq. 1.35 reduces to a form very similar to Eq. 1.26

$$V_p = \left(1 - \frac{p_B}{1 - p_A}\right)^{N-1} \cong 1 - (N-1)\frac{p_B}{1 - p_A}. \tag{1.37}$$

In short, spectral overlap alters the intensity of the unmodulated signal relative to the dipolar modulation and, if p_{AB} is not small, produces new modulation centered at $T = \tau$. Spectral overlap produces the same effects in 4pPELDOR [30]. The relative intensities in the PELDOR time trace are altered, and new dipolar modulation appears centered at $= 0$, τ_2, and $\tau_1 + \tau_2$ in addition to the expected dipolar modulation centered at $T = \tau_1$. It also can introduce nuclear modulation seen in the classical ESEEM experiments, that is unrelated to D.

1.5.1 Decreasing the Effect of Spectral Overlap

Spectral overlap is a problem in two situations in Eq. 1.35. One is when the partner spin in a pair has an EPR frequency close to ω_A so that $p_A > 0$ and it is excited by the observe pulse. There is no ideal way to minimize this effect. (1) Making measurements at higher frequency bands may help if there is significant g-anisotropy because that decreases the chance that both spins in a pair have similar EPR frequencies. (2) Decreasing p_A by decreasing the pulse turning angle seems a poor choice because the echo intensity decreases even faster than p_A. (3) Making the observe pulse more selective by increasing the pulse duration $t_{p,2}$ or using shaped pulses can make the partner spin less likely to fall within the frequency range excited by the observe pulse. This stratagem can decrease signal intensity and conflicts with another conventional requirement of Eq. 1.8, see Sect. 1.6. Fortunately, p_A affects only the unmodulated part of the PELDOR time trace and is not a major difficulty.

The second way that spectral overlap impacts PELDOR measurements is through p_{AB}, the probability that observe and pump pulses both excite the partner spin. Each mw pulse excites a band of frequencies, Fig. 1.9, which can overlap, producing a significant p_{AB}. This overlap does affect the modulation in the PELDOR time trace as seen in the last two terms in Eq. 1.35. The problem can be reduced by averaging PELDOR measurements for several values of τ so that the final term in Eq. 1.35 vanishes. It is also possible to decrease p_{AB} by adjusting the separation between observe and pump frequencies Δ, or by adjusting the excitation profiles of the pulses, e.g., using larger pulse widths or shaped pulses.

1.6 PELDOR with Finite Mw Pulse Duration

Earlier in Sects. 1.1–1.4, we considered an ideal PELDOR signal independent of experimental parameters, such as the pulse durations for observe (t_{p1}, t_{p2}) and pump (t_{p3}), resulting from the conventional assumptions of Eqs. 1.6–1.9. However, finite pulse durations do affect the PELDOR signal in typical experiments. In particular, PELDOR time traces from pairs at short distance have sharp features comparable to the mw pulse duration. These sharp features can be attenuated or lost completely. In addition, the dipole coupling, in such cases, can be comparable to the mw magnetic field strength, violating the assumption in Eq. 1.8.

1.6.1 Theoretical Approach

Understanding how the PELDOR time trace depends upon mw pulse duration requires careful consideration [11]. We will assume that Eqs. 1.6, 1.7, and 1.9 hold, and that $\Delta = |\omega_A - \omega_B| \gg |D|$. This lets us neglect the action of the mw magnetic fields on off-resonance spins, i.e., $p_{AB} = 0$ so that observe pulses at ω_A do not interact with B spins and pump pulses at ω_B do not interact with A spins. The time-dependent Hamiltonian of the spin pair has the form

$$\begin{aligned}\hat{H}/\hbar = {} & \omega_{0A}\hat{S}_{zA} + \omega_{0B}\hat{S}_{zB} + D\hat{S}_{zA}\hat{S}_{zB} \\ & + 2F_A(t)\omega_{1A}\cos(\omega_A t)\hat{S}_{xA} \\ & + 2F_B(t)\omega_{1B}\cos(\omega_B t + \varphi)\hat{S}_{x'B}\end{aligned} \tag{1.38}$$

where ω_{0Q} is the Zeeman frequency of the Q spin for $Q = A, B$; and $F_Q(t) = 1$ during the appropriate mw pulse and 0 at other times. In Eq. 1.38, the mw frequency can differ from the precise EPR frequency of a spin, e.g., $\delta_Q = (\omega_{0Q} - \omega_Q) \neq 0$. The mw magnetic fields are linearly polarized but their directions may differ and the PELDOR spectrometer uses coherent detection. In this situation, Eq. 1.38 is further simplified by passing into a special coordinate system, known as the two-frequency rotating frame or TFRF, where each spin rotates with its frequency ω_{0Q} around the laboratory axis z. This simplification gives the spin Hamiltonian

$$\begin{aligned}\hat{H}_{\mathrm{TFRF}} = {} & \delta_A\hat{S}_{zA} + \delta_B\hat{S}_{zB} + D\hat{S}_{zA}\hat{S}_{zB} \\ & + F_A(t)\omega_{1A}\hat{S}_{xA} + F_B(t)\omega_{1B}\hat{S}_{x'B}\end{aligned} \tag{1.39}$$

where the energy has units of angular frequency and the rapidly oscillating terms are neglected. The x and x' axes lie along the mw fields in the appropriate coordinate system. Prior to the mw pulses, the system is at thermal equilibrium with an equilibrium density matrix in the high-temperature limit of

$$\rho_0 \approx \frac{\hbar}{4kT}\left(\omega_A \hat{S}_{zA} + \omega_B \hat{S}_{zB}\right) \tag{1.40}$$

which holds above 0.5 K at X-band.

The PELDOR time trace $V(T)$ can be calculated from Eqs. 1.39–1.40 using special methods developed in [11], even when the mw pulse amplitudes are comparable to D. The experimental signals are described by four vectors which precess in their effective magnetic fields. For strong D, the PELDOR effect is observed when the mw fields affect both transitions of the Pake doublets.

The detailed calculations of $V(T)$ in [11] produce a rather simple equation for 3pPELDOR data when the entire echo is integrated. The $V(T)$ is the sum of constant and oscillating terms, similar to Eq. 1.5 [53, 54]

$$V(T) = v_0\left(v_\infty + \widetilde{K}(DT)\right) \tag{1.41}$$

where v_0 is the PELDOR signal intensity in the absence of the pump pulse (ignoring dynamic phase shifts [39]), $\widetilde{K}(DT)$ is the oscillating part of the PELDOR signal, and v_∞ is independent of T [11, 53]:

$$v_0 = f\left(\omega_A + \frac{D}{2}\right) + f\left(\omega_A - \frac{D}{2}\right), \tag{1.42}$$

$$\begin{aligned} v_\infty &= \left(\mu_{x+}\mu_{x-} + \mu_{y+}\mu_{y-}\right)\cos\left(\frac{Dt_{p3}}{2}\right) \\ &\quad + \left(\mu_{x-}\mu_{y+} - \mu_{x+}\mu_{y-}\right)\sin\left(\frac{Dt_{p3}}{2}\right), \end{aligned} \tag{1.43}$$

$$\widetilde{K}(DT) = \mu_{z+}\mu_{z-}\left(\eta_y \cos\left(DT + \frac{Dt_{p3}}{2}\right) - \eta_x \sin\left(DT + \frac{Dt_{p3}}{2}\right)\right). \tag{1.44}$$

The function, $f(\omega)$, describes the shape of the normalized EPR spectrum, and t_{p3} is the pump pulse duration. The components of vector $\vec{\mu}_\pm$ in Eqs. 1.43–1.44 are

$$\begin{aligned} \vec{\mu}_\pm &= \left(\mu_{x\pm}, \mu_{y\pm}, \mu_{z\pm}\right) \\ &= \left(\frac{\delta_{B\pm}}{\Omega_{B\pm}}\sin\left(\Omega_{B\pm}\frac{t_{p3}}{2}\right), \cos\left(\Omega_{B\pm}\frac{t_{p3}}{2}\right), \frac{\omega_{1B}}{\Omega_{B\pm}}\sin\left(\Omega_{B\pm}\frac{t_{p3}}{2}\right)\right) \end{aligned} \tag{1.45}$$

where $\delta_{B\pm} = (\omega_{0B} - \omega_B) \pm D/2$, and $\Omega_{B\pm} = \sqrt{\delta_{B\pm}^2 + \omega_{1B}^2}$. The η_x and η_y in Eq. 1.44 are

$$\eta_x = \frac{2d\sin\left(\sqrt{1+d^2}\frac{\omega_{1A}t_{p1}}{2}\right)^2}{(1+d^2)\sin(\omega_{1A}t_{p1})} \tag{1.46}$$

$$\eta_y = \frac{\sin\left(\sqrt{1+d^2}\omega_{1A}t_{p1}\right)}{\sqrt{1+d^2}\sin(\omega_{1A}t_{p1})} \tag{1.47}$$

$$d = \frac{D}{\omega_{1A}}. \tag{1.48}$$

The shape of $V(T)$ depends on the pump pulse duration t_{p3} through v_∞, $\widetilde{K}(DT)$, and $\vec{\mu}_\pm$. The observe pulse t_{p1} and t_{p2} influence $\widetilde{K}(DT)$ through Eqs. 1.46–1.47, as well as, the overall signal intensity [6, 11] through

$$\upsilon_0 = \frac{\pi\hbar\omega_A}{4kT}\sin(\omega_{1A}t_{p1})\sin\left(\frac{\omega_{1A}t_{p2}}{2}\right)^2. \tag{1.49}$$

The 3pPELDOR time trace in Eq. 1.41 converges to Eq. 1.5 as the pulse durations approach zero with constant turning angle and large Δ.

1.6.2 *Effect of Pulse Duration on the PELDOR Time Trace*

We can now compare experimental results on the effect of pump pulse width in Sect. 1.6.1 with the theoretical calculations. The 3pPELDOR time traces $\langle V(T)\rangle$ for randomly oriented pairs of spins with $J = 0$ were calculated, Fig. 1.10, for typical nitroxyl hyperfine A- and g-tensors: $A_{xx} = 0.4$, $A_{yy} = 0.7$ and $A_{zz} = 3.50$ mT; $g_{xx} = 2.0095$, $g_{yy} = 2.0061$ and $g_{zz} = 2.0025$. A Gaussian line shape with a first-derivative peak to peak width of 0.39 mT was used for all other sources of broadening.

The calculated time traces, Fig. 1.10, show that mw pulse duration has a remarkable effect on $V(T)$. Increases in pulse duration decrease the modulation amplitude and reduce $V(0)$. Despite these decreases, the frequency of the dipolar oscillations remains nearly constant for the same distance r. Thus, the distance between spins can be determined directly from the period of oscillations for any pulse duration.

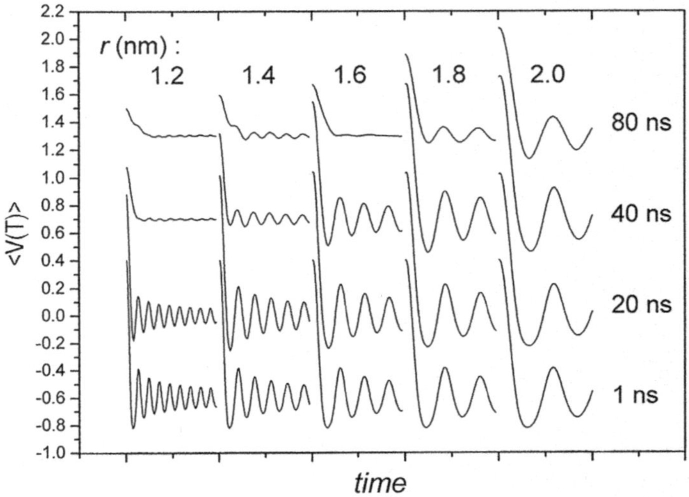

Fig. 1.10 The calculated PELDOR time traces for a set of distances r between spins and mw pulse durations with $t_{p1} = t_{p2} = t_{p3}$; the pulse turning angles for all the pulses are 120° and the spins have EPR parameters typical of nitroxyl radicals [53]

For $r < 2$ nm, the pulse durations should be less than ~20 ns while for $r > 2$ nm, the PELDOR signal is rather independent of pulse duration. These trends are seen in experiments, where an increasing the observe pulse durations from 30 to 90 ns leads to the complete disappearance of oscillations, Fig. 1.11.

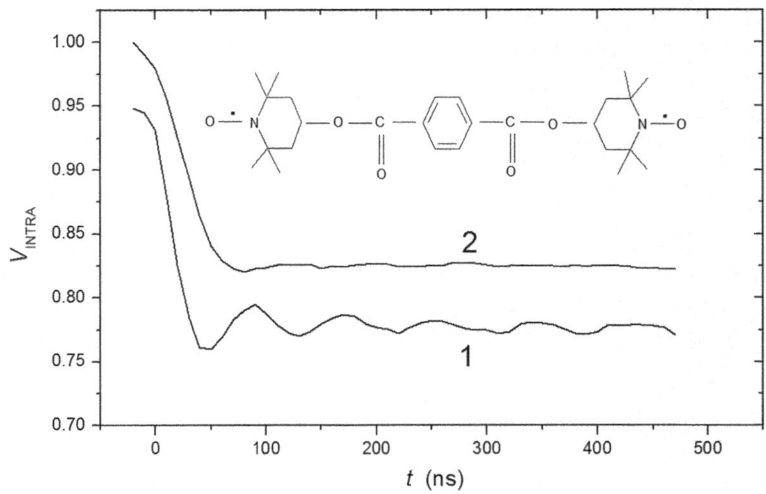

Fig. 1.11 Experimental PELDOR time traces for a frozen glassy biradical in d_8-toluene for pulse durations: *1* $t_{p1} = t_{p2} = 30$ ns, $t_{p3} = 40$ ns; *2* $t_{p1} = t_{p2} = 90$ ns, $t_{p3} = 40$ ns [53]

1.6.3 *Effect of Pulse Duration on* $F(r)$

In the absence of orientation selection, the time trace in Eq. 1.41 can be averaged over *r*:

$$V(T) = \int_{r_1}^{r_2} F(r)\langle v_0(v_\infty + \widetilde{K}(DT))\rangle dr \tag{1.50}$$

where $F(r)dr$ is the fraction of pairs having a distance in the interval r to $r + dr$; r_1 and r_2 are the limits of the possible distances, and $< \cdots >$ indicates the averaging over θ and the EPR spectrum of the B spins.

The normalized PELDOR time trace Eq. 1.31 for finite-width pulses has the form

$$V_n(T) = \int_{r_1}^{r_2} \widetilde{F}(r)K(r,T)dr, \tag{1.51}$$

$$K(r,T) = \langle v_0 \widetilde{K}(DT)\rangle, \tag{1.52}$$

$$\widetilde{K}(r) = \frac{F(r)}{\int_{r_1}^{r_2} F(r)\langle v_0(1 - v_\infty)\rangle dr}, \tag{1.53}$$

showing that Eq. 1.51 is a Fredholm integral equation of the first type. Its solution, giving the $\widetilde{F}(r)$ function, allows one to calculate the distance distribution function $F(r)$ as described in Sect. 1.3.

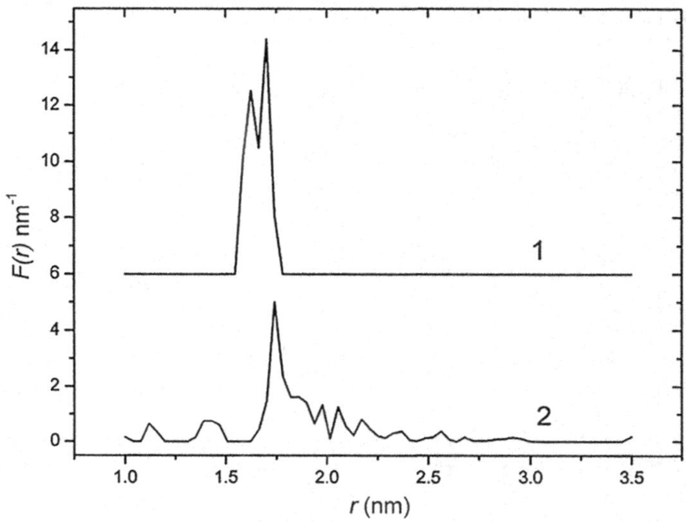

Fig. 1.12 The distance distribution functions for a biradical: *1* the effect of the mw pulse width was taken into account using Eq. 1.51; *2* the analysis assumed infinitely short mw pulses using Eq. 1.5 [53]

The $F(r)$ were extracted from the PELDOR time trace of a short biradical, with and without correcting for pulse duration [53]. The distance distribution functions obtained by two methods are similar, Fig. 1.12, and both gave $J = 1.4 \pm 0.5$ MHz. The distribution function based on Eq. 1.51 contains two maxima at distances of 1.62 and 1.7 nm, consistent with the two expected conformations of the radical. The distance distribution from the simpler Eq. 1.5 has one major peak, but significant intensity is spread among a broad range of peaks inconsistent with the biradical structure, Fig. 1.12. The effect of finite mw pulse duration in 4pPELDOR has also been examined [55].

1.7 Summary

This chapter has presented the basic concepts supporting the use of PELDOR as a method to study molecular and macromolecular structure on the nanometer scale. The next three chapters examine two important aspects of PELDOR measurements. The instrumentation used to make PELDOR measurements is described, along with its impact on the results. This is followed by an examination of some of the spins: free radicals and spin labels, that are commonly used as labels in chemical and biological studies using PELDOR. These chapters are followed by experimental examples that illustrate common approaches used in experimental applications of PELDOR in several fields.

References

1. Milov AD, Salikhov KM, Shirov MD (1981) Application of eldor in electron-spin echo for paramagnetic center space distribution in solids. Fiz Tverd Tela 23(4):975–982
2. Stein RA, Beth AH, Hustedt EJ (2015) A straightforward approach to the analysis of double electron-electron resonance data. Methods Enzymol 563:531–567. https://doi.org/10.1016/bs.mie.2015.07.031
3. Milov AD, Ponomarev AB, Tsvetkov YD (1984) Modulation beats of signal of double electron-electron resonance in spin-echo for biradical systems. J Struct Chem 25(5):710–713. https://doi.org/10.1007/Bf00747913
4. Klauder JR, Anderson PW (1962) Spectral diffusion decay in spin resonance experiments. Phys Rev 125(3):912–932. https://doi.org/10.1103/PhysRev.125.912
5. Hahn EL (1950) Spin echoes. Phys Rev 80(4):580–594
6. Salikhov KM, Semenov AG, Tsvetkov YD (1976) Electron spin echo and its applications. Nauka, Novosibirsk
7. Pannier M, Veit S, Godt A, Jeschke G, Spiess HW (2000) Dead-time free measurement of dipole-dipole interactions between electron spins. J Magn Reson 142(2):331–340. https://doi.org/10.1006/jmre.1999.1944
8. Kaplan DE, Hahn EL (1958) Experiences de double irradiation en resonance magnetique par la methode dimpulsions. J Phys-Paris 19(11):821–825. https://doi.org/10.1051/jphysrad:019580019011082100

9. Graf R, Demco DE, Gottwald J, Hafner S, Spiess HW (1997) Dipolar couplings and internuclear distances by double-quantum nuclear magnetic resonance spectroscopy of solids. J Chem Phys 106(3):885–895. https://doi.org/10.1063/1.473169
10. Emshwiller M, Hahn EL, Kaplan D (1960) Pulsed nuclear resonance spectroscopy. Phys Rev 118(2):414–424. https://doi.org/10.1103/PhysRev.118.414
11. Maryasov AG, Tsvetkov YD (2000) Formation of the pulsed electron-electron double resonance signal in the case of a finite amplitude of microwave fields. Appl Magn Reson 18 (4):583–605. https://doi.org/10.1007/Bf03162305
12. Milov AD, Maryasov AG, Tsvetkov YD (1998) Pulsed electron double resonance (PELDOR) and its applications in free-radicals research. Appl Magn Reson 15(1):107–143. https://doi.org/10.1007/Bf03161886
13. Tsvetkov YD (1989) ELDOR in ESE study of magnetic dipole-dipole interactions. In: Keijzers CP, Reijerse EJ, Schmidt J (eds) Pulsed EPR: a new field of applications. North Holland, Amsterdam, pp 206–218
14. Maryasov AG, Tsvetkov YD, Raap J (1998) Weakly coupled radical pairs in solids: ELDOR in ESE structure studies. Appl Magn Reson 14(1):101–113. https://doi.org/10.1007/Bf03162010
15. Parmon VN, Kokorin AI, Zhidomirov GM (1980) Stable biradicals. Nauka, Moscow
16. Bowman MK, Maryasov AG, Kim N, DeRose VJ (2004) Visualization of distance distribution from pulsed double electron-electron resonance data. Appl Magn Reson 26(1–2):23–39. https://doi.org/10.1007/Bf03166560
17. Abragam A (1961) The Principles of Nuclear Magnetism. Clarendon Press, Oxford
18. Raitsimring AM, Salikhov KM (1985) Electron spin echo method as used to analyze the spatial distribution of paramagnetic centers. Bull Magn Reson 7(4):184–217
19. Kutsovsky YE, Maryasov AG, Aristov YI, Parmon VN (1990) Electron-spin echo as a tool for investigation of surface-structure of finely dispersed fractal solids. React Kinet Catal L 42 (1):19–24. https://doi.org/10.1007/Bf02137612
20. Milov AD, Samoilova RI, Tsvetkov YD, Gusev VA, Formaggio F, Crisma M, Toniolo C, Raap J (2002) Spatial distribution of spin-labeled trichogin GA IV in the gram-positive bacterial cell membrane determined from PELDOR data. Appl Magn Reson 23(1):81–95. https://doi.org/10.1007/Bf03166186
21. Milov AD, Tsvetkov YD (1997) Double electron-electron resonance in electron spin echo: Conformations of spin-labeled poly-4-vinilpyridine in glassy solutions. Appl Magn Reson 12 (4):495–504
22. Uemura S, Okada M, Abedin KM, Nakatsuka H (1992) Fractal interpretation of Non-Lorentzian persistent hole shapes in organic glasses. Chem Phys Lett 189(2):193–196. https://doi.org/10.1016/0009-2614(92)85122-Q
23. Tsvetkov YD (2004) Peptide aggregation and conformation properties as studied by pulsed electron-electron double resonance. In: Bender C, Berliner LJ (eds) EPR: instrumental methods. Biological magnetic resonance, vol 21. Springer, New York. https://doi.org/10.1007/978-1-4419-8951-2
24. Milov AD, Maryasov AG, Tsvetkov YD, Raap J (1999) Pulsed ELDOR in spin-labeled polypeptides. Chem Phys Lett 303(1–2):135–143. https://doi.org/10.1016/S0009-2614(99)00220-1
25. Milov AD, Tsvetkov YD (2000) Charge effect on relative distance distribution of Fremy's radical ions in frozen glassy solution studied by PELDOR. Appl Magn Reson 18(2):217–226. https://doi.org/10.1007/Bf03162112
26. Milov AD, Ponomarev AB, Tsvetkov YD (1984) Electron electron double-resonance in electron-spin echo—model biradical systems and the sensitized photolysis of decalin. Chem Phys Lett 110(1):67–72. https://doi.org/10.1016/0009-2614(84)80148-7
27. Ponomarev AB, Milov AD, Tsvetkov YD (1988) Double electron-electron resonance in electron-spin echo and the spatial distribution of radicals formed by irradiation of frozen cyclohexane. Khim Fiz 7(12):1673–1679

28. Ponomarev AB, Milov AD, Tsvetkov YD (1990) Double electron-electron resonance in electron-spin echo—spatial distribution of radicals forming during radiolysis of polyethylene, monocarboxylic and dicarboxylic acids. Khim Fiz 9(4):498–503
29. Salikhov KM, Khairuzhdinov IT, Zaripov RB (2014) Three-pulse ELDOR theory revisited. Appl Magn Reson 45(6):573–619. https://doi.org/10.1007/s00723-014-0541-7
30. Salikhov KM, Khairuzhdinov IT (2014) Four-pulse ELDOR theory of the spin ½ label pairs extended to overlapping EPR spectra and to overlapping pump and observer excitation bands. Appl Magn Reson 46(1):67–83. https://doi.org/10.1007/s00723-014-0609-4
31. Edwards TH, Stoll S (2018) Optimal Tikhonov regularization for DEER spectroscopy. J Magn Reson 288:58–68. https://doi.org/10.1016/j.jmr.2018.01.021
32. Jeschke G, Koch A, Jonas U, Godt A (2002) Direct conversion of EPR dipolar time evolution data to distance distributions. J Magn Reson 155(1):72–82. https://doi.org/10.1006/jmre.2001.2498
33. Milov AD, Tsvetkov YD, Formaggio F, Oancea S, Toniolo C, Raap J (2003) Aggregation of spin labeled trichogin GA IV dimers: distance distribution between spin labels in frozen solutions by PELDOR data. J Phys Chem B 107(49):13719–13727. https://doi.org/10.1021/jp035057x
34. Chiang YW, Borbat PP, Freed JH (2005) The determination of pair distance distributions by pulsed ESR using Tikhonov regularization. J Magn Reson 172(2):279–295. https://doi.org/10.1016/j.jmr.2004.10.012
35. Jeschke G, Panek G, Godt A, Bender A, Paulsen H (2004) Data analysis procedures for pulse ELDOR measurements of broad distance distributions. Appl Magn Reson 26(1–2):223–244. https://doi.org/10.1007/Bf03166574
36. Dzuba SA (2016) The determination of pair-distance distribution by double electron-electron resonance: regularization by the length of distance discretization with Monte Carlo calculations. J Magn Reson 269:113–119. https://doi.org/10.1016/j.jmr.2016.06.001
37. Matveeva AG, Nekrasov VM, Maryasov AG (2017) Analytical solution of the PELDOR inverse problem using the integral Mellin transform. Phys Chem Chem Phys 19(48):32381–32388. https://doi.org/10.1039/c7cp04059h
38. Tikhonov AN, Arsenin VY (1977) Solutions of ill-posed problems. Scripta series in mathematics. Winston; Halsted Press, Washington, New York
39. Bowman MK, Maryasov AG (2007) Dynamic phase shifts in nanoscale distance measurements by double electron electron resonance (DEER). J Magn Reson 185(2):270–282. https://doi.org/10.1016/j.jmr.2006.12.011
40. Jeschke G, Chechik V, Ionita P, Godt A, Zimmermann H, Banham J, Timmel CR, Hilger D, Jung H (2006) DeerAnalysis2006—a comprehensive software package for analyzing pulsed ELDOR data. Appl Magn Reson 30(3–4):473–498. https://doi.org/10.1007/Bf03166213
41. Borbat PP, Freed JH (2002) Double-quantum ESR and distance measurements. In: Berliner LJ, Eaton SS, Eaton GR (eds) Distance measurements in biological systems by EPR, vol 19. Springer, New York. https://doi.org/10.1007/b111467
42. Godt A, Schulte M, Zimmermann H, Jeschke G (2006) How flexible are poly(para-phenyleneethynylene)s? Angew Chem Int Edit 45(45):7560–7564. https://doi.org/10.1002/anie.200602807
43. Sicoli G, Mathis G, Aci-Seche S, Saint-Pierre C, Boulard Y, Gasparutto D, Gambarelli S (2009) Lesion-induced DNA weak structural changes detected by pulsed EPR spectroscopy combined with site-directed spin labelling. Nucleic Acids Res 37(10):3165–3176. https://doi.org/10.1093/nar/gkp165
44. Schnegg A, Dubinskii AA, Fuchs MR, Grishin YA, Kirilina EP, Lubitz W, Plato M, Savitsky A, Mobius K (2007) High-field EPR, ENDOR and ELDOR on bacterial photosynthetic reaction centers. Appl Magn Reson 31(1–2):59–98
45. Mobius K, Lubitz W, Savitsky A (2013) High-field EPR on membrane proteins—crossing the gap to NMR. Prog Nucl Mag Res Sp 75:1–49. https://doi.org/10.1016/j.pnmrs.2013.07.002

46. Savitsky A, Dubinskii AA, Flores M, Lubitz W, Mobius K (2007) Orientation-resolving pulsed electron dipolar high-field EPR spectroscopy on disordered solids: I. Structure of spin-correlated radical pairs in bacterial photosynthetic reaction centers. J Phys Chem B 111 (22):6245–6262. https://doi.org/10.1021/jp070016c
47. Denysenkov VP, Prisner TF, Stubbe J, Bennati M (2006) High-field pulsed electron-electron double resonance spectroscopy to determine the orientation of the tyrosyl radicals in ribonucleotide reductase. P Natl Acad Sci USA 103(36):13386–13390. https://doi.org/10.1073/pnas.0605851103
48. Polyhach Y, Godt A, Bauer C, Jeschke G (2007) Spin pair geometry revealed by high-field DEER in the presence of conformational distributions. J Magn Reson 185(1):118–129. https://doi.org/10.1016/j.jmr.2006.11.012
49. Bowen AM, Tait CE, Timmel CR, Harmer JR (2013) Orientation-selective deer using rigid spin labels, cofactors, metals, and clusters. In: Timmel CR, Harmer JR (eds) Structural information from spin-labels and intrinsic paramagnetic centres in the biosciences. Springer, Berlin, Heidelberg, pp 283–327. https://doi.org/10.1007/430_2013_115
50. Yang ZY, Kise D, Saxena S (2010) An approach towards the measurement of nanometer range distances based on Cu^{2+} ions and ESR. J Phys Chem B 114(18):6165–6174. https://doi.org/10.1021/jp911637s
51. Lovett JE, Bowen AM, Timmel CR, Jones MW, Dilworth JR, Caprotti D, Bell SG, Wong LL, Harmer J (2009) Structural information from orientationally selective DEER spectroscopy. Phys Chem Chem Phys 11(31):6840–6848. https://doi.org/10.1039/b907010a
52. Bode BE, Plackmeyer J, Prisner TF, Schiemann O (2008) PELDOR measurements on a nitroxide-labeled Cu(II) porphyrin: Orientation selection, spin-density distribution, and conformational flexibility. J Phys Chem A 112(23):5064–5073. https://doi.org/10.1021/jp710504k
53. Milov AD, Naumov BD, Tsvetkov YD (2004) The effect of microwave pulse duration on the distance distribution function between spin labels obtained by PELDOR data analysis. Appl Magn Reson 26(4):587–599. https://doi.org/10.1007/Bf03166585
54. Milov AD, Tsvetkov YD, Maryasov AG, Gobbo M, Prinzivalli C, De Zotti M, Formaggio F, Toniolo C (2012) Conformational properties of the spin-labeled tylopeptin B and heptaibin peptaibiotics based on PELDOR spectroscopy data. Appl Magn Reson 44(4):495–508. https://doi.org/10.1007/s00723-012-0402-1
55. Milov AD, Grishin YA, Dzuba SA, Tsvetkov YD (2011) Effect of pumping pulse duration on echo signal amplitude in four-pulse PELDOR. Appl Magn Reson 41(1):59–67. https://doi.org/10.1007/s00723-011-0232-6

Chapter 2
Experimental Techniques

Abstract PELDOR or DEER, like most forms of spectroscopy, always involves tradeoffs between sensitivity and resolution. Some tradeoffs are made during the instrumental design of the spectrometers that measure PELDOR spectra, and other tradeoffs are made in the experimental design of the measurements. This chapter considers the instrument design first. It examines the capabilities needed to make PELDOR measurements and then examines the designs of several spectrometers operating at different mw frequencies. This is followed by a discussion of how capabilities of the spectrometer and the operating and measurement parameters affect sensitivity and resolution in PELDOR distance distribution spectra.

2.1 Basic Spectrometer Functions

The PELDOR effect is the change in the ESE signal when pump pulses are applied to the spin system. An ESE spectrometer is necessary to generate and measure the echo signal, but some additional equipment is needed to produce the pump pulses and to apply them to the sample. ESE and PELDOR spectrometers operate in rather narrow mw frequency ranges with $\omega/2\pi$ scattered between 3 and 600 GHz [1–10]. Several different conventions break the microwave (mw) frequency range into a series of bands designated by letters and these bands are used to indicate the approximate operating frequency of spectrometers. A list of some of the more commonly-used designations appear in Table 2.1.

An ESE spectrometer provides much of the core functionality for PELDOR measurements, but requires some additional capabilities. We consider the general functions needed for PELDOR and then look in Sects. 2.2–2.4 at ways several spectrometers implement these requirements.

© Springer Nature Switzerland AG 2019

Y. D. Tsvetkov et al., *Pulsed Electron–Electron Double Resonance*,
https://doi.org/10.1007/978-3-030-05372-7_2

Table 2.1 Common mw band designations

Band	Frequency $\omega/2\pi$ (GHz)	Wavelength (cm)
L	1–2	15–30
S	2–4	7.5–15
C	4–8	3.75–7.5
X	8–12	2.5–3.75
K_u	12–18	1.67–2.5
K	18–26.5	1.13–1.67
K_a	26.5–40	0.5–1.13
Q	33–50	0.6–0.9
mm	40–300	0.1–0.5
V	40–75	0.4–0.6
W	75–110	0.27–0.4
D	110–170	0.18–0.27

2.1.1 The ESE Core

Several ESE spectrometers have been described in detailed [6, 11–13]. In this section, we focus on the specific features needed for PELDOR measurements. Three different approaches are in use at EPR frequencies through W-band. They are summarized in the simplified block diagrams for a basic PELDOR spectrometer, Fig. 2.1. Each design has, at its core, a standard ESE spectrometer which generates an ESE at ω_A. The performance of this core is one major factor that determines sensitivity, and consequently, the range of distances that can be measured by PELDOR techniques. Therefore, design principles that optimize sensitivity of an ESE spectrometer are also vital for PELDOR measurements.

A low-power mw source on the left side of each block diagram, Fig. 2.1a–c, supplies the mw frequency for the mw pulses and for the signal detector, via a reference arm. The pulse former unit converts the low-power mw output of the source into weak pulses which are amplified by a mw power amplifier and delivered to a mw resonator containing the sample. The ESE goes from the sample in the resonator to the detector where it is converted from a short mw burst into the outline, or envelope, of that burst, known as a video signal. The signal processor block at the right side of the block diagram captures that signal and sends it to a computer where the signals are assembled into a time trace. The pulse programmer block controls and orchestrates all these functions.

2.1.2 MW Source

The mw source for the observe frequency ω_A is usually a voltage-controlled oscillator (VCO), a Gunn diode oscillator with phase-locked loop (PLL) stabilization, or a synthesizer. The ESE is not affected by the phase noise of the mw source to the

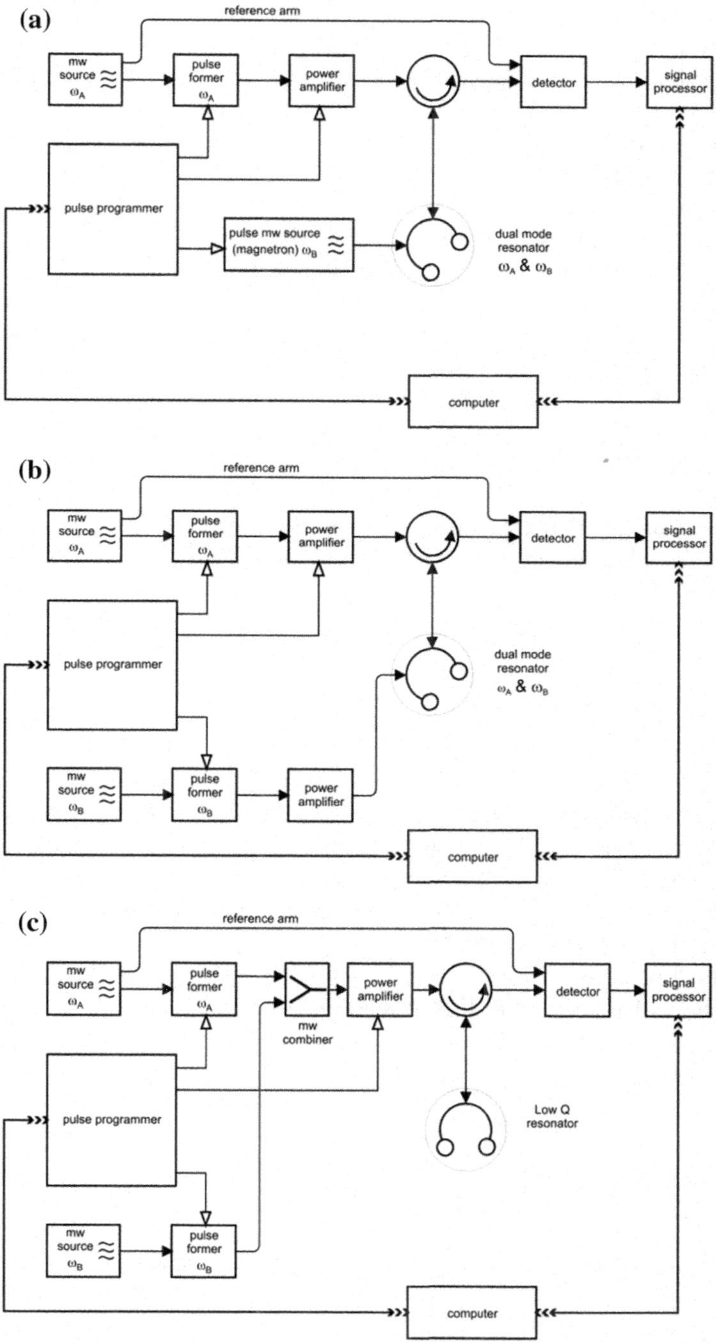

Fig. 2.1 Block diagrams of some X- and Q-band PELDOR spectrometers: **a** bimodal resonator and pulse magnetron as the pump mw source [14]; **b** bimodal resonator and two TWTAs; **c** low Q, overcoupled resonator in most modern spectrometers

extent that the CW EPR signal is, but long-term frequency drift is a problem for signal averaging. There is generally a very large difference between the length of the reference arm and the effective length of the signal-generating arm, particularly when the large propagation delay through a traveling wave tube amplifier (TWTA) is considered. Any difference in effective path length converts frequency drift of the source into a drift in the detected signal phase from the mixer, because the mw mixer is a phase-sensitive detector.

Phase drift during a PELDOR measurement subtly alters the shape of the PELDOR time trace and can impact the distance distribution. But, frequency stability is only one factor in spectrometer phase drift. Two other important factors are resonance-frequency changes in the sample resonator caused by microphonics, sample movement or thermal drift; and voltage regulation of the mw amplifiers, particularly the TWTA accelerating voltage.

Currently, mw oscillators operating at the spectrometer frequency can be used through Q-band. However, a low-frequency source at $\omega/2\pi \approx 6\text{–}7\,\text{GHz}$ that is stabilized by a dielectric resonator (DRO) typically is used with subsequent frequency mixing and multiplication for spectrometers operating above Q-band. The DRO serves as the master oscillator, providing low phase and amplitude noise to preserve spectrometer sensitivity.

2.1.3 Pulse Former Unit

The pulse former uses the frequency of the mw source to produce weak mw pulses. The pulse former can be as simple as a mw switch that simply gates the mw output of the source on and off. Most pulse formers also provide control of the amplitude and phase of the pulses, while a few designs produce very complex shaped pulses, chirped pulses or composite pulses. However it is accomplished, it is far easier to form mw pulses at low power and then amplify them than it is to form the pulses from a high power source.

The mw pulses used in PELDOR measurements are fairly short in duration, so that the rising and falling edges can be a significant fraction of the total length, which can impact the pulse excitation spectrum.

2.1.4 Power Amplifier

The mw power amplifier amplifies the low-power mw pulses to the high power levels needed to excite the sample. This amplifier is generally a gated, or pulse, amplifier, meaning that it is in an off state most of the time and is turned on, so that it amplifies, only when needed. A gated amplifier generally consumes much less power than a CW amplifier; power is required only while it is gated on, a few percent of the time. This means it has a smaller, more reliable power supply; is much cheaper; dissipates much less power; and requires much less cooling. For high-precision measurements where stability is vital, the duty cycle of the power amplifier is best kept constant, to keep the power consumption and temperature of the amplifier constant.

A gated amplifier produces virtually no noise in the off state. An operating amplifier is producing noise. The noise it generates is usually measured in terms of the equivalent thermal noise at the input. So, for example, an amplifier with a 30 dB gain (1000-fold power amplification) and a noise figure of 3 dB (equivalent noise power at the input 2-fold, i.e., 3 dB, higher than thermal noise) would produce noise at its output with 33 dB or 2000 times more power than the thermal noise level at room temperature. The PELDOR signal is often weaker than the room temperature thermal noise level, so noise from the power amplifier must be negligible at the time of the signal. This usually is achievable with amplifiers designed for fast turn-off, however, the amplifier turn off is one contributing factor to the spectrometer dead time.

Gating the amplifier is not a good method to produce short mw pulses. The rise and fall times of the amplifier output are generally much longer than those of a pulse former, and would produce short pulses that have rather triangular or trapezoidal shapes. In addition, there are usually very large phase shifts in the amplified mw output during the rise and fall of the output, resulting in unintended and poorly characterized shaped pulsed that are far from optimal for EPR spectroscopy. If the power amplifier is operated somewhat beyond it linear amplification range in order to obtain the highest power, the pulse shape and phase is subtly altered. Usually the edges become sharper, so pulses are squarer after the power amplifier than at the input.

The power amplifier is usually operated by turning it on; letting it reach a stable operating state; sending one or more mw pulses into it for amplification; and then turning it off. The amplifier may be gated on several times during a single measurement, e.g., once for the first two pulses in a 4pPELDOR measurement, again for the pump pulse, and a final time for the third observe pulse.

2.1.5 Pulse Programmer

A multifunction pulse programmer controls the mw pulses that generate the ESE signal. It controls: the time intervals between the pulses; the pulse width, phase and shape from the pulse formers; and gating of the mw power amplifier. Those factors control when the signal occurs. So the pulse programmer also must control signal acquisition because signal acquisition must be coherently synchronized with signal generation. Signal acquisition involves any gating needed to protect the receiver from damage or overload, and the measurement or digitization of a precise portion of the detected ω_A signal. If signal generation and acquisition use more than one timebase, it is important that they be synchronized and coherent with each other to avoid subtle sources of noise. In PELDOR experiments, the pulse programmer must additionally control the pump pulse at ω_B.

2.1.6 Resonator

The mw pulse sequence at ω_A is fed through a circulator to a resonator containing the sample under investigation. The signal, along with any ringing or reflections from the resonator, comes back to the circulator and is directed to the receiver. The resonator is located inside the computer-controlled magnet, not shown in Fig. 2.1. An iron-core magnet is typical at X- and Q-band and a superconducting cryomagnet at W-band. The resonator supports mw standing waves at ω_A and ω_B. Usually, the sample is placed where the mw electric field is zero, and the mw magnetic field is maximum for both mw frequencies and perpendicular to the field of the surrounding magnet. Resonators are discussed in greater detail in Sect. 2.5.

2.1.7 Detector and Signal Processor

The detector converts the mw burst that is the ESE into a broadband signal centered at zero frequency (homodyne detection), or centered at a lower intermediate frequency ω_{IF} (heterodyne detection). This usually is done by mixing the signal with a mw frequency from the reference arm, ω_A for homodyne detection, or $\omega_A \pm \omega_{IF}$ for heterodyne detection. The resulting signal is then converted into some digital form by the signal processor.

The detector input is usually protected from the power amplifier noise, from the mw pulses themselves, and from ringing of the resonator by a mw limiter or switch that blocks the detector input when appropriate. This protection is needed because too much power can damage or destroy the detector, or can saturate the detector so that its output is distorted. A saturated detector can take quite a long time, microseconds or milliseconds, to recover.

An X-band detector has a low-noise mw preamplifier before the mixer and video amplifiers after the mixer for maximum sensitivity. Detectors at higher mw frequencies are restricted by the limited performance and availability of components and the ESE signal may arrive at the mixer without preamplification.

2.1.8 The PELDOR Arm

In PELDOR experiments, the sample must be exposed to pump mw pulses at ω_B, in the time interval between the observe pulses at ω_A, Figs. 1.1 or 1.2. The pump pulses are often amplified by the same power amplifier as the observe pulses, Fig. 2.2c. Several variations used at X-band are described in Sect. 2.2.

The PELDOR arm contains a mw source for ω_B. It could be derived from ω_A by frequency shifting, but that is not necessary. The PELDOR signal does not depend strongly on the exact frequency or phase of ω_B, and does not require a stable, low-noise mw source. This permits a cheap, non-coherent oscillator to be used for ω_B.

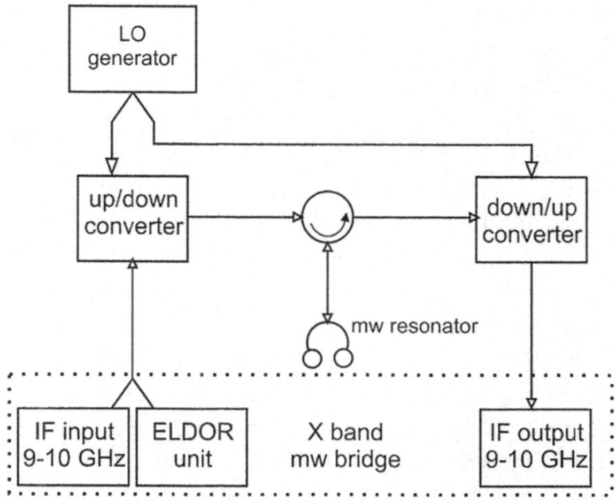

Fig. 2.2 Simplified scheme of the Bruker BioSpin intermediate frequency (IF) PELDOR spectrometer concept [25, 26]; the operating frequency band is defined by an auxiliary low-power local oscillator LO, but the pulse sequence preparation and the signal detection is carried out at X-band

2.2 X- and Q-Band Instruments

Three implementations, at X- and Q-bands, of the basic design just discussed in Sect. 2.1 are shown in Fig. 2.1. The multiple pulses at ω_A and ω_B are fed into the resonator two different ways. One design uses a resonator with separate inputs for each frequency. This requires an independent amplifier for high-power pump pulses at ω_B, e.g., a magnetron or an independent TWTA, to feed directly into the resonator through the "pump" waveguide, Fig. 2.1a, b [15–17].

The other method, Fig. 2.1c, uses only one power amplifier. The low-power pulses at ω_A and ω_B are combined as a single input to the high-power amplifier and the output is fed into the resonator. In this case, a single-mode, low-Q resonator is usually used; while the first method allows use of a single-mode or a bi-modal resonator. A single-mode resonator is easy to set up and tune, but has poor sensitivity when the frequency difference $\Delta\omega_{AB}$ is large because the resonator Q, and therefore sensitivity, must be reduced so that both ω_A and ω_B fall within its bandwidth [15, 18]. High-Q, bimodal resonators achieve good sensitivity [19–22], but require additional tuning, and additional design care to ensure that the mw magnetic fields of both modes have strong, uniform overlap at the sample, see Sect. 2.6.4.

2.2.1 The ICKC Spectrometer

PELDOR spectrometers operating at X-band are currently the most common. The X-band PELDOR spectrometer based on a coherent pulse EPR spectrometer [14], Fig. 2.1a, is examined first. This ESE spectrometer was used for many years to develop and test PELDOR methods at the Voevodsky Institute of Chemical

Kinetics and Combustion (ICKC) in Novosibirsk, Russia. This spectrometer introduced several novel features when it was built, including: independent sources and amplifiers for very intense mw pulses at ω_A and ω_B; and the ability to modulate, or cycle, the phase of the mw observe pulses.

The mw pulses at ω_A can have a power as high as 10 kW with duration $t_p \geq$ 20 ns. These pulses are produced by two-stage amplification of the weak output of a stabilized klystron, which also feeds the reference arm. The weak pulses are amplified to 1 kW by a TWTA. These 1 kW pulses could be used directly as the observe pulses to generate ESE signals or they could be amplified to ~10 kW by an injection-locked magnetron which preserves the phase and frequency of the original low-power pulses. This pulse amplification process means that the phase of the high-power mw pulses is controlled at low-power, where it is easily done. This ability to control pulse phase was used to implement phase cycling for the first time in ESE spectroscopy, which reduced the dead time for the ESE signal to ~200 ns and eliminated unwanted responses, such as the FID and the resonator ringing [14, 23, 24].

The pulse programmer generates pulses for external synchronization and controls the detector. The pulse duration and the intervals between pulses are discretely controlled with 1 ns increments. The intervals between pulses can extend to 10 ms and can be swept. The programmer also produces a control pulse for the phase shifter, which changes the phase of the first pulse in an ESE sequence. The pulse repetition rate lies between $2 * 10^2$ and 10^4 Hz. The receiver noise factor is 5–6 dB. Digital acquisition of the ESE signal is possible. The magnetic field was controlled by the NMR signal of water protons to an accuracy of 3 ppm.

This ESE spectrometer [14] served as the core for a PELDOR spectrometer. A magnetron operating in a self-excitation mode at a fixed frequency of $\omega_B/2\pi =$ 9400 MHz provided the pump pulses. The magnetron acts as a combined mw source, pulse former and power amplifier, controlled by pulse modulation of its anode voltage. The maximum mw power output of the magnetron is 5 kW with $t_p = 30–40$ ns.

The spectrometer can generate ESE signals from the A or the B spins at the same external magnetic field B_0. To do this, the magnetron gives a pair of pulses at ω_B, forming an ESE signal from B spins at ω_B. That ESE signal is amplified by a mw amplifier, detected by a simple mw diode detector and observed on an oscilloscope. This feature provides a simple, experimental determination of the rotation angle θ of the B spins.

2.2.2 The E580 Spectrometer

Many researchers use the E580 X-band spectrometer from the Bruker BioSpin ELEXYS series for PELDOR measurements, or the earlier ESP 380 model [25–27]. A low-power mw source, or "ELDOR Unit", is added to the main mw bridge of the basic E580 to provide ω_B, as in Fig. 2.1c. The $\omega_B/2\pi$ of this ELDOR Unit can be set in the range of 9–10 GHz in 100-Hz steps. The $\omega_A/2\pi$ is typically ~9.8 GHz, making the maximum difference $\Delta\omega_{AB}/2\pi$ about 800 MHz.

Pulses at ω_A and ω_B are formed by high-speed switches, which provide mw pulses with rise and fall times of 1 ns and a minimum duration of 4–8 ns. Pulses at both ω_A and ω_B are amplified by a single TWTA with a maximum output power of ~1 kW. The amplitude, duration, and phase of the ω_A pulses are specified independently from those at ω_B. The amplified pulses are fed into a single-mode, overcoupled dielectric or split-ring resonator. A quadrature mixer preceded by a low-noise preamplifier detects the signals.

The fact that pulses at both frequencies are formed independently, but amplified by the same TWTA, is an important difference between the E580 and the spectrometer developed by the ICKC [14]. This has meant that only single-mode, broadband resonators are used with the E580, while the ICKC spectrometer [14] uses a bimodal rectangular resonator with two waveguide inputs: one for pulses at ω_A from the TWTA, and the other for pulses at ω_B from the magnetron.

2.3 High-Frequency Instruments

PELDOR spectrometers at frequencies above X-band face practical and technical limitations. In particular, bimodal resonator design becomes increasingly difficult at high frequencies. This restricts the frequency difference $\Delta\omega_{AB}$ to the resonator bandwidth.

EPR spectra of unoriented paramagnetic centers become considerably broader at frequencies above X-band due to g-factor anisotropy. Consequently, the PELDOR modulation is considerably weaker for comparable pump pulses at frequencies beyond X-band, because the degree of excitation of the spectrum p_B decreases. However, these high-frequency limitations greatly improve orientation selection.

2.3.1 The IF Approach

The X-band E580 spectrometer of Bruker BioSpin [25, 26] can be modified for use in the L-, S-, Q-, and W bands for CW and pulse EPR, including PELDOR [26, 27]. These modifications use the Intermediate Frequency (IF) concept for operating frequencies outside the X-band, Fig. 2.2. The most significant advantage of this approach is that proven X-band technology is used for excitation and detection regardless of the operating EPR frequency.

A pulsed spectrometer operating at both the X- and Ku-bands was developed at Cornell University mainly for double quantum coherence (DQC) spectroscopy [28]. This spectrometer uses pulses with an excitation bandwidth of ~70 G or ~2π *

200 MHz and can excite the whole EPR spectrum of nitroxyl labels for DQC experiments.

A homodyne double-channel X-band bridge forms pulses with a duration of 3–5 ns at four fixed phases with a minimum separation in time of 5 ns. This bridge applies the IF concept to operate at Ku band, $\sim 2\pi*17$ GHz. A pump pulse module allows PELDOR measurements at both bands. The frequency difference $\Delta\omega_{AB}/2\pi$ can be selected in the range of 0–700 MHz.

2.3.2 W-Band Instruments

A homodyne W-band spectrometer [29] similar to Fig. 2.1c or Fig. 6 of [30], has two independent mw sources based on frequency multiplication. The tunable source consists of an $\sim 2\pi*7.3$-GHz master oscillator and a $\times 13$ frequency multiplier, giving $\omega_B/2\pi$ in the range of 94.9 ± 1.3 GHz with steps of 13 kHz. A similar source has a fixed frequency for $\omega_A/2\pi = 94.9\,\text{GHz}$. The pulses are formed by independent pulse formers with high-frequency *p–i–n* diode switches and they are combined as the input to a solid-state mw power amplifier with up to 1 W output power. The pulse power at the input of the TE_{100} cylindrical resonator is estimated to be 300–350 mW [29]. With a loaded Q-value, $Q_L \sim$ 1000, the π pulse has a duration $t_p \sim 26\,\text{ns}$. The spectrometer dead time is ~ 50 ns with the resonator at room temperature and ~ 150 ns at low temperature because conductivity of the resonator walls and, therefore, the Q-value, increase.

A top-of-the-line W-band spectrometer, Fig. 2.3, was developed at the Free University of Berlin [31–33]. The cryomagnet has a 114 mm diameter warm bore, to accommodate a helium cryostat with a useful diameter of ~ 90 mm. Field scans of up to 1000 G are performed by auxiliary superconducting sweep coils. The mw bridge is based on a heterodyne circuit with an intermediate frequency of $2\pi*4$ GHz. Three DRO master oscillators with similar frequencies of $\sim 2\pi*7$ GHz are available to help track the natural frequency of the resonator.

After multiplication by $\times$ 13, ~ 10 mW of the mw power at $\omega_A/2\pi = 95\,\text{GHz}$ is used to produce the local oscillator $\omega_{LO}/2\pi = 99\,\text{GHz}$ for the detector; the rest passes through a system of pulse-forming switches and phase shifters to the input of an IMPATT diode amplifier with ~ 300 mW output power, and then through an attenuator and circulator to a cylindrical TE_{011} or Fabry–Perot resonator.

The EPR signals pass back through the circulator and a *p–i–n* switch which protects the receiver from high-power reflected mw pulses. The signal is mixed down to $\omega_{IF}/2\pi = 4\,\text{GHz}$; amplified by a LNA; converted to a video signal by the quadrature detector; and the components of the EPR signal are fed into the signal processor.

The dead time of the spectrometer in pulse mode is ~ 20 ns, and the noise figure of the mw receiver is 11 dB or better. PELDOR results from this spectrometer on the geometry and distances in photosynthetic reactive centers are discussed in Sect. 7.2.2.

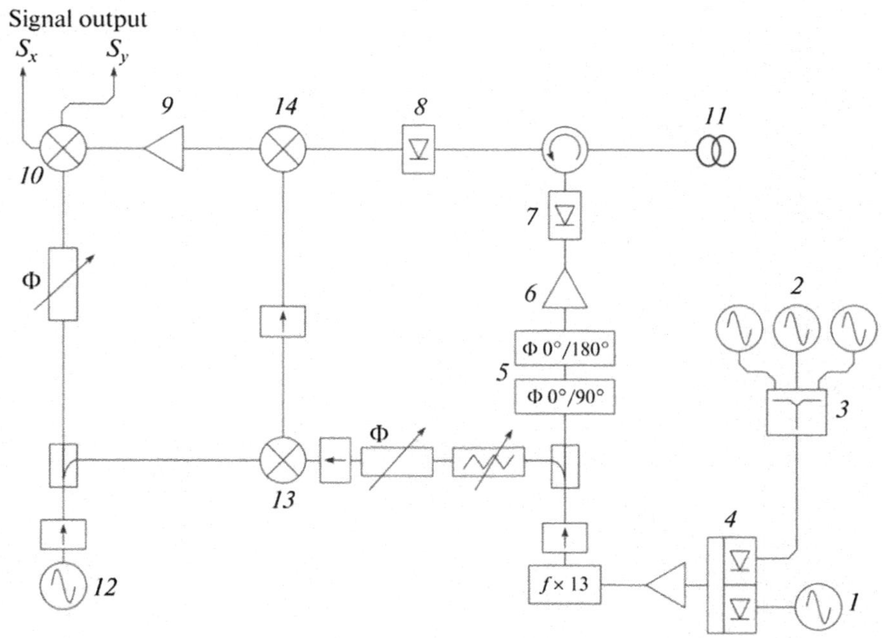

Fig. 2.3 Block diagram of a pulsed 95 GHz heterodyne EPR spectrometer [30]: *1* oscillator at $\omega/2\pi = 7.3 \pm 0.023\,\text{GHz}$; *2* oscillators at 7.3 + 0.23, 7.3, and 7.3–0.23 GHz; *3* power combiner; *4, 7* fast *p–i–n* switches; *5* fast phase modulators; *6* avalanche-transit diode power amplifier; *8* protective fast *p–i–n* diode switch; *9* a 4 GHz LNA amplifier; *10* mixer–quadrature detector; *11* TE011 resonator; *12* a 4 GHz oscillator; *13* mixer to generate the local oscillator $\omega_{LO}/2\pi = 4 + 95 = 99\,\text{GHz}$; and *14* mixer to shift the signal to $\omega_{IF}/2\pi = 99 - 95 = 4\,\text{GHz}$

2.3.3 180 GHz PELDOR

Several laboratories have designed PELDOR spectrometers operating above W-band. One is a heterodyne pulse EPR spectrometer operating at $\omega_A/2\pi \approx 180\,\text{GHz}$, Fig. 2.4 [34–36]. The spectrometer generates the mw pulse sequence at $\sim 2\pi$*45 GHz and then doubles the frequency twice to obtain $\omega_A/2\pi \approx \omega_B/2\pi \approx 180\,\text{GHz}$.

Pulses of $\sim$100 ns duration are formed by switches in two independent channels and combined by a magic-T coupler. The mw pulse sequence is frequency doubled, amplified, doubled again. Waveguide circulators are not available at this band, so a quasi-optical circulator delivers $\sim$20 mW to a cylindrical resonator with an inner diameter of 2.2 mm via low-loss corrugated oversized waveguide.

An elliptical mirror focuses the ESE signal from the resonator into the receiver through a conical waveguide taper. The cryogenic magnet of the spectrometer creates a static magnetic field of up to 7.0 T with a sweep range of 1.5 kg. This spectrometer was used to investigate the orientation selection in pairs of tyrosine radicals in proteins [37, 38].

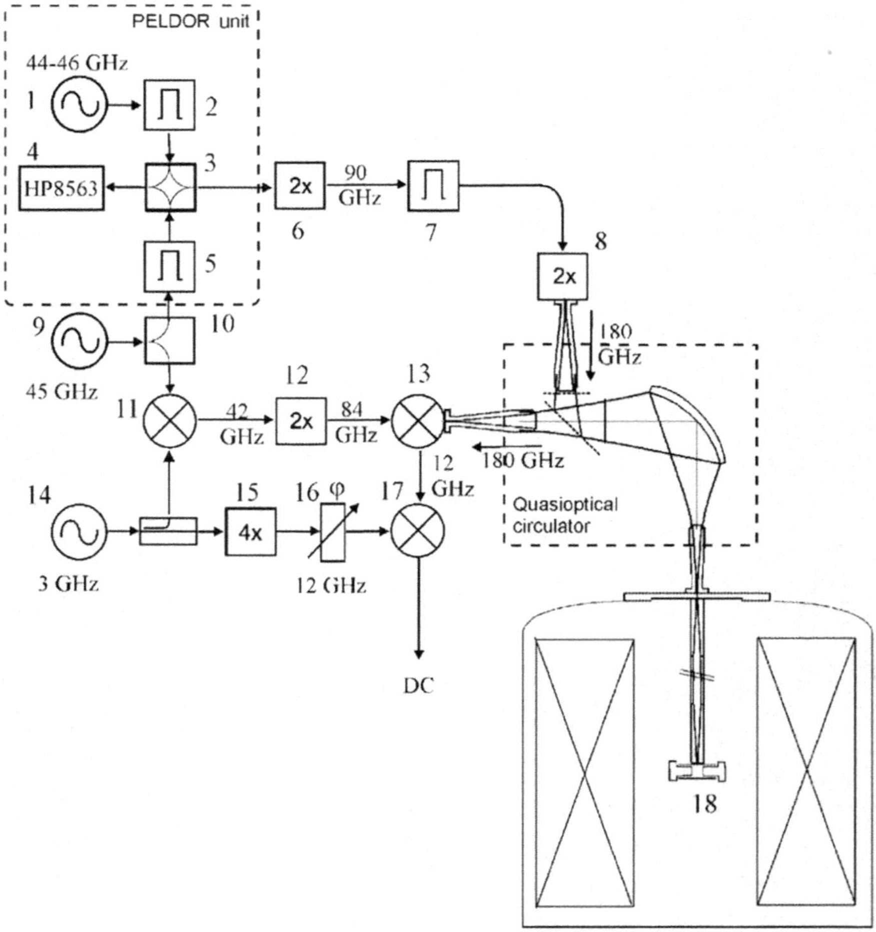

Fig. 2.4 Block diagram of a 180 GHz pulse EPR/ELDOR spectrometer: *1* variable frequency oscillator; *2, 5, 7* fast switches; *3* magic-T; *4* spectrum analyzer; *6, 8, 12* frequency doublers; *9, 14* oscillators; *10* power divider; *11, 13, 17* mixers; *15* frequency multiplier; *16* phase shifter; *18* resonator; 45 GHz amplifiers are omitted for clarity. Reproduced from Hertel et al. [35], with permission of John Wiley and Sons, copyright 2005

2.3.4 263 GHz PELDOR

The record high $\omega_A/2\pi \approx 263\,\text{GHz}$ quasi-optical commercial spectrometer was designed by Bruker BioSpin. The BRUKER E780 instrument [10, 39] performs CW and pulse EPR experiments at magnetic fields close to 9.4 T. The mw bridge is based on a heterodyne IF design consisting of two units: an X-band transmitter/ receiver and the 263 GHz quasi-optical front-end. This minimizes the number of millimeter-wave components by performing the crucial mw generation and signal detection at X-band. The X-band transmitter includes two mw sources.

A TE_{011} single-mode resonator is coupled to the quasi-optical front-end via a mode converter and corrugated waveguide, which substantially reduces mw losses to below 1 dB. The resonator Q-value can reach 1000 and, with the mw power of 15 mW at the resonator, the optimal $\pi/2$ pulse duration is 50–55 ns. Since the size of the resonator scales with mw wavelength, $\lambda \sim 1$ mm, the sample tube has outer and inner diameters as small as 0.33 and 0.2 mm, respectively, with a sample volume $\sim$30 nL for a 1 mm long sample.

Impressive spectral and orientation resolution is achieved. Specifically, at $B_0 \sim 9.6$ T, the g-factor resolution is 30-fold better than at X-band and $\sim$3 better than W-band. This gives much higher orientation selection for PELDOR.

However, PELDOR experiments at 263 GHz are more difficult to implement and execute [10]. The limited power and increased mw losses require long pulses, giving weaker EPR signals, while g-anisotropy produces broader EPR spectra. One expects weaker modulation but better orientation selection than at lower frequencies. The high absolute sensitivity of the HF-EPR spectrometer allows PELDOR measurements on samples with limited volumes, particularly promising for rare biological/biochemical materials.

2.4 B_0-Jump PELDOR

The PELDOR spectrometers described above use two mw sources for ω_A and ω_B. An alternative approach is to rapidly change the magnetic field from $B_0 = \omega_A/2\pi\gamma$, by ΔB_0, corresponding to a pump frequency $\omega_B = \gamma(B_0 + \Delta B_0)$. PELDOR or ESE experiments based on this technique were proposed and conducted to measure spin–lattice relaxation, spectral diffusion [40], and transfer of saturation in EPR spectra, and for studying slow molecular motions [41–43].

The PELDOR method with B_0 jump was used to determine distances between spins in biradical **3-1** and in a related biradical [44]. The three-pulse stimulated ESE, Fig. 2.5, enabled detection of the signal. A magnetic field jump equivalent to the pump frequency ($\Delta\omega_{AB} = \gamma\Delta B$) was applied between the second and third pulses of the three pulse stimulated ESE sequence, Fig. 2.5a or a field sweep provided very broad pump excitation, Fig. 2.5b. The stimulated ESE was measured as a function of the delay τ between the first and second pulses.

A Bruker BioSpin ESP-380 pulse spectrometer was used with a rectangular H_{102} resonator modified with walls of thin 25 μm foil of nonmagnetic stainless steel that was "semi-transparent" to the magnetic field pulse [44]. The magnetic field pulse was produced by low-inductance coils replacing the modulation coils. The low mw conductivity of the walls gave the low Q required for PELDOR experiments. The pulse current source creating the field jump contains corrective RC-circuits to produce the flattest possible field pulse with minimum rise and fall times. The shift of the EPR line in the ESE-detected spectrum was used to determine that the rise and fall times of the magnetic field pulse were under 1 μs. The maximum magnetic

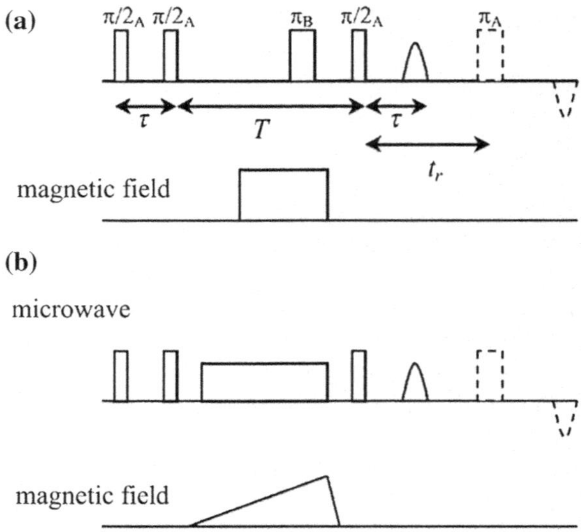

Fig. 2.5 Pulse sequences for field-step ELDOR: the three-pulse stimulated ESE sequence $\pi/2_A - \tau - \pi/2_A - T - \pi/2_A - \tau - echo$ with the pump pulse π_B during the field jump. **a** Pumping with fixed magnetic field jump; **b** scanning pumping; to overcome the dead time problem, the ESE may be refocused by an additional π_A pulse (*dashed line*). Reproduced from Kulik et al. [44] with permission of Elsevier Limited, copyright 2002

field pulse amplitude was ±170 G, and the pulse duration could be set from 1.0 μs to 1.0 s [44].

A similar setup for magnetic field jump PELDOR experiments has been used at W-band [32]. The advantage of field jumps over the double-frequency PELDOR method is quite clear at W-band: it permits use of a high-Q resonator to obtain the highest-possible sensitivity. PELDOR using a magnetic field jump seems very promising; it does not require a second mw source for the pump pulse, which becomes expensive at higher frequencies. The pulsed field version of PELDOR shares all the advantages of existing single-frequency EPR methods.

2.5 Resonators: Construction, Q, and Dead Time

The resonator is located inside the EPR magnet—an iron-core magnet for X- and Q-band, or a superconducting cryomagnet for W-band. The resonator has several mw standing waves or modes. Usually, the frequencies of only one mode, for a single-mode resonator, or two modes, for a bimodal resonator, fall in the frequency range of the spectrometer. Ideally, the sample is placed where the mw electric field is zero and the mw magnetic field is maximum and perpendicular to the field of the surrounding magnet.

In PELDOR experiments, the sample must be exposed to pump pulses at ω_B, in the time interval between the pulses at frequency ω_A, Figs. 1.1 and 1.2. Two ways are used to feed the multiple pulses at ω_A and ω_B into the resonator depending on whether it is a single-mode or a bimodal resonator, Sect. 2.2.

A single-mode, low-Q resonator is relatively easy to design and build. It has only one active mode, so the bandwidth must be large to support the mw fields at both ω_A and ω_B. Single-mode resonators usually have only one input and are said to operate in reflection because the signal travels in the opposite direction from the resonator from the observe pulses. However, single mode resonators with two inputs/outputs have been used and operate in transmission. Because the same resonator mode is used for both frequencies, the mw magnetic fields of both frequencies have excellent spatial overlap, which is important for PELDOR. A single-mode resonator is easy to tune and operate, but has poor sensitivity when the frequency difference $\Delta\omega_{AB}$ is large because the resonator Q, and therefore sensitivity, must be reduced so that both ω_A and ω_B fall within the resonator bandwidth.

Bimodal resonators achieve good sensitivity by using separate high-Q modes for ω_A and ω_B [3, 19–22], but introduce several difficulties. Bimodal resonators suitable for PELDOR are difficult to design because: the two modes must be tunable to the desired ω_A and ω_B; the mw magnetic field maximum of each mode must uniformly overlap each other and the sample; and mw magnetic fields of each mode must be perpendicular to the field of the EPR magnet. In addition, an input specific for each mode is usually desired. It is possible to excite both modes from a single input, but requires a method of adjusting the coupling to each mode. At high mw frequencies, bimodal resonators require very exacting mechanical tolerances and adjustments, making them difficult to construct and use. New approaches to bimodal resonator design do show promise [22].

2.5.1 Resonator Properties

The resonator for PELDOR experiments has two conflicting requirements. First, it must create large mw B_1 fields at the sample to excite a large portion of the EPR spectrum. For instance, to flip spins at $g = 2.0$ by a π turning angle, a 10-ns duration pulse must have a mw field amplitude of $B_1 \sim 10$ G. Second, the resonator must excite the sample at two frequencies ω_A and ω_B. These two requirements are difficult to satisfy simultaneously.

In CW EPR spectrometers, the B_1 in a resonator depends on its Q_{Loaded} and the mw input power P:

$$\begin{aligned} B_1 &= \delta(Q_{Loaded}P)^{1/2} \\ Q_{Loaded} &= \frac{\omega_0}{\Delta\omega} \end{aligned} \tag{2.1}$$

where δ is a parameter depending on the resonator design and sample position; ω_0 and $\Delta\omega$ are the resonant frequency and the resonator bandwidth, respectively.

The mw power incident on the resonator input is usually limited in a PELDOR spectrometer; at X-band, to ~ 1 kW. The simplest method to increase B_1 is to increase the resonator Q, as it is usually done in CW EPR spectroscopy; but this is unacceptable for pulse EPR spectroscopy, since a high Q factor increases the dead time due to resonator ringing and distorts the mw pulse shape. To avoid distorting short, rectangular pulses, the band width of the resonator must satisfy the condition [4]

$$\frac{\Delta\omega}{2\pi} \geq \frac{0.57}{t_p} \tag{2.2}$$

where t_p is the pulse duration, limiting the Q factor to

$$Q_{Opt} \leq 1.75 t_p \frac{\omega_0}{2\pi}. \tag{2.3}$$

This means, at X-band, that the Q factor for undistorted 10 ns rectangular pulses is $Q_{Opt} \leq 175$.

The dead time is the interval after a pulse during which the signal cannot be measured. Many factors contribute to the dead time including: unwanted signals, such as the FID; mismatches and reflections in the mw circuit; noise from power amplifiers; recovery of amplifiers from overload; and the "ringing" of the resonator after the last mw pulse. The resonator ringing must decay to roughly the level of the noise power, which takes a time t_R, illustrated in Fig. 2.6.

The dead time is an important consideration in ESE spectroscopy because the ESE intensity at small τ is important. This is not the case in 3p- and 4pPELDOR measurements which typically use fairly large values of $\tau > t_R$, with

$$t_R = \frac{Q}{2\omega_0} \ln\left(\frac{P_0}{P_n}\right) \tag{2.4}$$

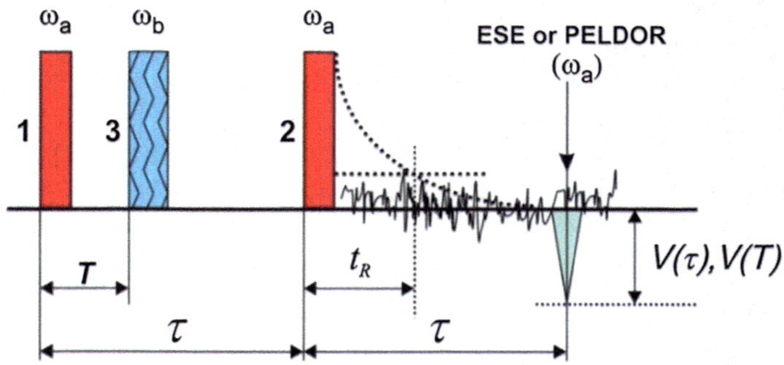

Fig. 2.6 Dead time or ringdown time in 3pPELDOR

where P_0 is the mw pulse power and $P_n = kT\omega_0/2Q$ is the noise power. Estimates for $P_0 = 1\,\text{kW}$, $Q = 500$, $T = 300\,\text{K}$, and $\omega_0/2\pi = 9\,\text{GHz}$, give $t_R = 120\,\text{ns}$. A detailed analysis was presented in [11, 32, 45].

If it is necessary to reduce t_R, Eq. (2.4) shows that low-Q resonators must be used. However, this reduces sensitivity proportional to $Q^{1/2}$. A number of methods to reduce the dead time from ringing are used in pulse EPR spectroscopy, including: mw delay lines [38, 46, 47], mw pulse phase cycling [14], remote or refocused ESE [48, 49], bimodal resonators [15–17], loop-gap resonators [50], and detection of longitudinal magnetization [51].

2.5.2 *Single-Mode Resonators*

The Q-value must be low if a single-mode resonator is used for PELDOR experiments:

$$Q \leq \frac{\omega_0}{\Delta\omega_{AB}}. \tag{2.5}$$

With nitroxyl labels and organic radicals, $\Delta\omega_{AB}/2\pi$ can be $\geq$ 80–100 MHz at X-band, limiting Q to 100–130. For single-mode resonators, the competing need for a high B_1 and a low Q-value can be satisfied only by replacing standard cavity resonators with special resonant structures.

There are several methods to achieve the B_1 required in a single-mode resonator. One is to use semi-lumped structures in which the electric and magnetic mw fields are spatially separated, e.g., resonant helices [52], slotted-tube resonators [53], loop-gap resonators [50, 54], and split-ring resonators [15]. Such resonators have a much smaller volume, which increases B_1 and largely compensates for the loss of sensitivity from the low Q-value. The pioneering CW electron double resonance work used a helical resonator [55] and a rectangular bimodal resonant cavity [19].

The low-Q dielectric resonator ER 4118X-MD-5 from Bruker BioSpin [56] is widely used at X-band. Its dielectric resonator is an alumina tube made from a single crystal of Al_2O_3. It has good sensitivity, provides reasonable B_1 at moderate mw power, and its resonant frequency changes only a few MHz with temperature. But, it is extremely difficult to make alumina that is absolutely free of Cr(III) ions and even a trace amount produces rather strong signals in EPR spectra. These impurity signals are temperature dependent and can complicate data analysis and interpretation of weak samples. ESE signals from the impurity Cr(III) signal is generally seen only near 4 K for very short values of τ and is not a common problem for PELDOR measurements of nitroxyl labels and other organic radicals. Replacement of the standard alumina tube by a bismuth germinate, $Bi_4(GeO_4)_3$ or BGO, tube reduces the CW EPR impurity signals and the resonant frequency changes very little with temperature [57].

A cylindrical TE_{011} mode dielectric resonator at K_a-band with $\omega_A/2\pi = 33-34$ GHz was constructed for PELDOR measurements at cryogenic temperatures [18]. The estimated Q_{Loaded} was about 180, pulse duration was 10 ns for pump and 17 ns for observe pulses at $\Delta\omega_{AB}/2\pi \sim 180$ MHz.

A very effective method of increasing the bandwidth of the resonator and lowering the resonator Q-value is to increase its coupling to the waveguide or transmission line, i.e., overcoupling, so $\beta > 1$. Adjustment of the coupling [25, 58] changes Q_{Loaded}, $\Delta\omega$, and B_1. Resonance curves for dielectric resonators and split-ring resonators used by Bruker Biospin in PELDOR experiments are presented in [15]. For these resonators at $\Delta\omega_{AB}/2\pi = 500$ (dielectric resonator) and 700 MHz (split-ring), field strength $\gamma B_1/2\pi$ may be as high as 90 and 60 MHz, depending on the extent of overcoupling.

2.5.3 Bimodal Resonators

Crossed bimodal cavity resonators have found wide application in CW double resonance spectroscopy [59]. Bimodal resonators with $\omega_A = \omega_B$ but having mutually-orthogonal mw B_1 fields at the sample are used in pulse EPR spectroscopy to "uncouple" signal excitation from detection and reduce spectrometer dead time. The isolation between modes may be as high as 70–80 dB. The mw pulses in one resonator mode that excite the ESE signal are substantially attenuated in the other mode for detection. As a result, devices to protect the detector are no longer needed, reducing the dead time for an ESE signal to 30 ns. Bimodal resonators can be tuned for PELDOR with one mode resonant at ω_A and the other at ω_B, giving good sensitivity and large B_1 for $\Delta\omega_{AB}/2\pi = 700$ MHz or more.

The crossed bimodal cavity resonator used in PELDOR studies with the ICKC spectrometer is similar in design to the resonator proposed for CW double resonance experiments [19]. Both modes of the resonator are tuned by introducing quartz or Teflon rods in the appropriate volume. The coupling between the modes is -20 to -30 dB at $\omega_A/2\pi = 9.4$ GHz with $Q_A = 220$ and $Q_B = 150$. The $\Delta\omega_{AB}/2\pi$ may be set as large as 150 MHz.

At W-band, a double-cylindrical cavity was designed with two mw modes TE_{011} and TE_{012} rather close in frequency [22]. The sample is placed along the axis of the collinear cylinders. This is a novel arrangement for bimodal resonators for PELDOR because the mw B_1 fields of both modes are coaxial, not crossed. The quality factor is 4400 for the TE_{012} mode and 2900 for the TE_{011} mode. The mw B_1 fields measured by nutation at only 400 mW of mw power are 3.1 G and 2.9 G, respectively. The frequency difference $\Delta\omega_{AB}/2\pi$ between the TE_{012} and TE_{011} modes may be adjusted from $\sim$80 to $\sim$660 MHz for orientation selection experiments.

2.6 Performance Considerations

2.6.1 *Sensitivity*

The sensitivity in PELDOR experiments, in terms of the minimum number of particles detected, is governed by the sensitivity of the base ESE spectrometer. The initial amplitude of the PELDOR signal at a fixed τ corresponds to the ESE signal amplitude. Calculations of ESE spectrometer sensitivities have been made many times, e.g., see [1, 4, 11, 59]. One way to express the sensitivity is as the signal-to-noise ratio relative to the unavoidable noise from random thermal fluctuations in the spectrometer. For excitation of the full spectrum, the signal-to-noise ratio, *S*/*N*, for pulse EPR spectrometers depends on the operating frequency as [17, 31]:

$$(S/N)_0 \propto \left[\frac{Q_0}{V_0 F \Delta kT}\right]^{1/2} \frac{\omega_0^n}{kT} \tag{2.6}$$

where V_0 is the effective resonator volume, *F* is the noise figure of the receiver, Δ is the receiver bandwidth, *k* is the Boltzmann constant, and *T* is temperature. The exponent *n* depends on specific experimental conditions and can even change sign: $7/2 \leq n \leq -3/2$. For instance, for a small sample with a constant number of spins, n = 3/2 [60].

The minimal detectable number of spins for an "ideal" ESE spectrometer is [4]:

$$N_{min} = \frac{4\gamma kT}{g^2 \beta^2 T_2^* \omega_0^2} \sqrt{\frac{2kT_B V_0 T_1}{0.79 \pi t_R}} \tag{2.7}$$

where *g* is the electron g-factor, β the Bohr magneton, t_R the measurement time, and T_1, T_2^* are relaxation times. This formula assumes that the two pulses can be placed close enough in time that ESE decay is negligible.

As a rule, sensitivity increases with frequency. For example, under similar experimental conditions, sensitivity at X-band appeared to be 30–40 times higher than at S-band [61]. The sensitivity of Q-band spectrometers was $10-10^3$ times higher than that of X-band spectrometers, depending on the experimental conditions [62]. A comprehensive analysis of ESE spectrometer sensitivity was reported in [17].

Sensitivity can be increased by signal averaging. The efficiency depends on the signal detection method used: coherent, i.e., phase-sensitive, or incoherent. For coherent detection, sensitivity increases with the square root of the number of averages $\sqrt{N}$, while for incoherent detection, sensitivity increases more slowly as $4\sqrt{N}$ [4]. The optimum pulse repetition rate for signal averaging is [4]

$$\upsilon_{opt} = \frac{0.8}{T_1} \tag{2.8}$$

but only if the rate does not vary. Pauses or fluctuations in the pulse rate, e.g., for data transfers or computer overhead, can produce large spikes known as 'T_1' noise that does not average efficiently.

The sensitivity of an ideal coherent X-band ESE spectrometer is about 10^{10} – 10^{11} spins [4], provided that the spectrum is fully excited and the relaxation times were $T_1 = T_2 = 10^{-6}$s. This corresponds to the sensitivity of state-of-the-art CW EPR spectrometers.

The actual sensitivity of a pulse spectrometer is considerably worse for several reasons. The first is non-optimal design of the resonator. Short duration pump pulses and ESE signals require large resonator bandwidth and a low Q-value, leading to a small signal and a small S/N ratio. The second reason is non-optimal repetition rate for the observe pulses, and, finally, incomplete excitation caused by large spectral width or inhomogeneity of B_1 over the sample. The influence of all these factors was thoroughly discussed in [1, 4]. As a rule, the sensitivity of pulse spectrometers is 10–100 times lower than the sensitivity of CW instruments. PELDOR measurements can typically be made with about 100 μM spin label concentration in 50 μL of sample by commercial X-band spectrometers.

A characteristic of PELDOR experiments is that they are performed with a fixed time τ_0 between the ω_A pulses that produce the ESE signal. The τ_0 is generally not extremely short, since this would limit the time interval T and truncate the PELDOR time trace, causing a loss of distance information. Consequently, considerable phase relaxation and decay of the ESE signal occur. The PELDOR signal intensity (measurement sensitivity) decreases at large τ_0 by phase relaxation characterized by a time T_f, so that:

$$S/N \sim (S/N)_0 \exp\left(-\left(\frac{\tau_0}{T_f}\right)^n\right) \qquad (2.9)$$

with n between 1 and 2 [4, 63]. The PELDOR information is only a small fraction of the ESE signal and is proportional to p_B, further decreasing the PELDOR S/N ratio.

The sensitivity in PELDOR experiments is several times lower than in ordinary pulse EPR experiments. Experience suggests that a sample of spin-labeled protein and other biopolymer should contain $\geq 10^{12}$ molecules for a reliable distance distribution measurement.

However, there is considerable opportunity to improve the experimental PELDOR sensitivity by: deuterating the matrix and protein [64, 65], varying times T and τ during the measurement [66], measuring at higher mw frequencies [67–69], exploiting light-induced electron spin polarization [70, 71], using alternative pulse sequences [13, 42], improving the spin labels [72–76], employing bimodal resonators [22, 77], stitching together measurements from complementary 3p and 4p pulse sequences [78], and optimizing the tuning of pump and observe pulses [79].

2.6.2 Signal Processing Improvements

Sensitivity improvements are not limited just to increasing the signal. Removal of noise from the measurement also increases the S/N ratio. This is seen with phase cycling to remove resonator ringing and FIDs from the PELDOR time traces. Recently, a promising technique for increasing PELDOR sensitivity using a wavelet transform was demonstrated [80–82]. This basically removes noise from the time trace that does not have the characteristics of a PELDOR signal. Reduction of signal averaging times for time-domain signals by as much as two orders of magnitude was shown, while retaining the fidelity of the underlying signals. Excellent signal recovery was possible when the initial noisy signal has an $S/N \gtrsim 3$. The two order of magnitude reduction in signal averaging reported corresponds to a ten-fold sensitivity increase.

2.6.3 PELDOR Distance Range

The range of distances r that can be measured from oscillations of the PELDOR signal is limited by the range of times T in the PELDOR time trace. Roughly, a half-cycle of an oscillation is needed to determine its frequency accurately. The PELDOR signal oscillations occur at the dipolar frequency $\omega_D = \gamma^2\hbar/r^3$. Thus, the minimum dipole frequency, or maximum distance r_{max}, that can be measured, is limited by the maximum time interval T_{max} in the PELDOR time trace. Because the PELDOR signal is measured from a two-pulse ESE, T_{max} must be less than the interval τ_0 between detection pulses. These considerations set a distance limit of

$$r_{max} \approx \left[\frac{\gamma^2\hbar}{\pi}\tau_0\right]^{1/3}. \tag{2.10}$$

The two-pulse ESE signal decay sets a practical limit on τ_0. The ESE intensity decreases as τ increases in a process known as phase relaxation, characterized by a time constant T_f. Phase relaxation has a complex dependence on the concentration of electron and nuclear spins, temperature, molecular motion and other properties of the spin system. In PELDOR experiments, the high concentrations of protons, e.g., in proteins, organics or frozen aqueous solutions, commonly set an upper limit on T_f. Spin diffusion among the proton spins drives spectral diffusion of the electron spins, and phase relaxation, even at temperatures below 4 K [4, 63]. Proton spin diffusion causes the ESE to relax at large τ roughly as $\exp\left[-(\tau/T_f)^2\right]$, with $T_f \sim 2-5$ μs for typical organic or aqueous matrices. The value of $\tau_0 \sim 5$ μs gives a reasonable benchmark of $r_{max} \sim 8$ nm. Deuteration of the matrix can increase T_f, making measurements at $\tau_0 > 50$ μs feasible [64, 65], for $r_{max} \sim 16$ nm.

The classic PELDOR modulation, Eq. 1.10, was derived for uniform excitation of each Pake doublet, which are roughly $2\omega_D$ wide. Thus, uniform excitation requires $\gamma B_1 > 2\omega_D$ for both the pump and the observe pulses, or $\pi > 2\omega_D t_\pi$ for both ω_A and ω_B where t_π is the duration of π pulses. This sets a minimum distance of

$$r_{min} = \left[2\frac{\gamma^2\hbar}{\pi}t_\pi\right]^{1/3} \tag{2.11}$$

for $t_\pi \sim 30$ ns, $r_{min} \sim 1.2$ nm. Small exchange interactions between radicals becomes comparable to ω_D at this distance, see Sect. 3.4, complicating measurement of the distance.

These rough estimates indicate that the range of distances measurable from oscillations of the PELDOR time trace is 1.2–8.0 nm. In practice, the range seems to be more like $r_{min} \geq 1.6$ nm and $r_{max} \leq 8$ nm, although this range can be enlarged in some situations. The distance distribution $F(r)$ between spins can be determined reliably over this entire range.

2.6.4 Measurement of B_1

In pulse EPR spectroscopy, it is important that the spectrometer be adjusted so that mw pulses turn the spin magnetization through specified angles θ. This is done by adjusting the amplitude B_1 or duration t_p of the mw pulses. For example, in PELDOR experiments, the pump pulse at ω_B usually has a π turning angle and the observe pulses at ω_A have $\pi/2$ and π turning angles.

The B_1 must be determined not at its maximum in the resonator, but at the site of the sample under investigation. The easiest way to do this is to measure the nutation signal, or Tory oscillations, for a sample with a narrow EPR line [15, 83, 84]. The full spectral line must be excited and the sample must have a long phase relaxation time.

Coal is a convenient and frequently-used solid sample for determining B_1, but the EPR characteristics of coal vary with its source and its exposure to air. Solutions of stable radicals with EPR lines that are narrow in comparison with B_1, e.g., Fremy's salt **5-2**, are also used [83]. In these nutation experiments, the signal amplitude is modulated at the frequency, $\omega_1 = \gamma B_1$, permitting direct measurement of B_1.

Another convenient approach allows B_1 to be determined using the PELDOR sample in the resonator. This method is applicable both for narrow and broad EPR spectra. The shape of the ESE from two, equal-width mw pulses is recorded while varying the amplitude of the pulses at the peak of the EPR spectrum. The ESE signal, for incoherent or 'power' detection, is a symmetric double peak with a dip at the center when $\theta = \pi$ for either full or partial excitation of the spectrum [1, 4, 30]. With modern spectrometers, the in-phase signal is antisymmetric with positive and negative lobes of equal amplitude and shape.

Once θ is known for a specific set of conditions, it may be adjusted to any desired value using a calibrated mw attenuator between the power amplifier and the resonator to set the mw power going into the resonator. This method is far more accurate than adjusting the pulse width, because the nominal pulse width is often different from the actual width, and because the rise and fall times of the mw field in the resonator are not negligible.

The value of B_1 depends on the mw pulse frequency relative to resonator frequency. With a single mode resonator, at least one mw frequency differs from that of the resonator in a PELDOR experiment, Fig. 2.7a, and its B_1 is reduced. This off-resonant reduction is in addition to the B_1 reduction caused by the lower Q-value needed to accommodate pulses at ω_A and ω_B. A bimodal resonator lets mw pulses at ω_A and ω_B be placed at the center of their own mode, which allows both modes to have better Q-values, for even greater B_1 fields, Fig. 2.7b.

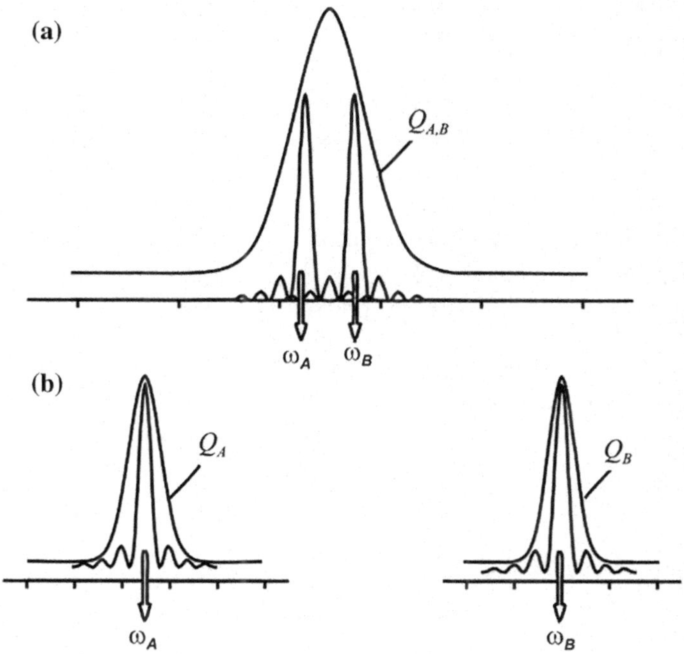

Fig. 2.7 Locations of observe pulses at ω_A and the pump pulses at ω_B, relative to the resonance curve for **a** a single-mode and **b** a bimodal resonator [30]

2.6.5 Determination of p_B

The EPR spectrum, the desired $\Delta\omega_{AB}$, and the duration of the mw pulses limit choices for ω_A and ω_B in PELDOR. The excitation spectra of the observe and pump pulses should not overlap; i.e., it is necessary that $\Delta\omega_{AB}t_p \gg 1$; but, at the same time, $\Delta\omega_{AB} < \varDelta$, where $\varDelta$ is the effective width of the EPR spectrum. For nitroxyl labels at X-band, $\varDelta \sim 300$ MHz; an experiment with $\Delta\omega_{AB} \sim 100$ MHz requires $t_p \gtrsim 40$ ns, which is easily met. However, there are inherent limitations for PELDOR with narrow EPR spectra or with short pulses.

The parameter p_B in Eq. (2.12) is necessary to analyze the PELDOR time trace. It is related to the rotation angle θ of spins B under the ω_B pulse. If the width of the spectrum of spins B is small compared to the pump pulse B_1, all spins rotate by the same angle and p_B can be defined as

$$p_B = \sin\left(\frac{\theta}{2}\right)^2 = \sin\left(\frac{\gamma B_1 t_p}{2}\right)^2 \tag{2.12}$$

For a broad EPR spectrum, (2.12) should be averaged over the EPR line shape $g(\omega)$. For a rectangular pulse, p_B can be calculated as [1, 4]:

$$p_B = \int_{-\infty}^{+\infty} \frac{\omega_1^2\, g(\omega)}{\omega_1^2 + (\omega_B - \omega)^2} \frac{\sin\left(\frac{t_p}{2}\sqrt{\omega_1^2 + (\omega_B - \omega)^2}\right)^2}{2\pi} d\omega \tag{2.13}$$

where $\omega_1 = \gamma B_1$. The p_B value can be calculated if the EPR line shape $g(\omega)$ and the pump pulse duration t_p are already known.

Experimental determination of p_B seems more reliable. For example, p_B can be determined from the phase relaxation of a model system with a known concentration of uniformly distributed spins using Eq. 1.17. This can be done with ease for glassy solutions of stable radicals. The p_B value can also be obtained for radical pairs known to have $N = 2$ using Eq. 1.16. Methods to determine p_B can be used to determine p_A.

References

1. Mims WB (1965) Electron echo methods in spin resonance spectrometry. Rev Sci Instrum 36 (10):1472–1479. https://doi.org/10.1063/1.1719359
2. Blumberg WE, Mims WB, Zuckerman D (1973) Electron-spin echo envelope spectrometry. Rev Sci Instrum 44(5):546–555. https://doi.org/10.1063/1.1686179
3. Huisjen M, Hyde JS (1974) Pulsed EPR spectrometer. Rev Sci Instrum 45(5):669–675. https://doi.org/10.1063/1.1686710
4. Salikhov KM, Semenov AG, Tsvetkov YD (1976) Electron spin echo and its applications. Nauka, Novosibirsk

5. Holczer K, Schmalbein D (1987) ESP 380: a high-power, general-purpose pulsed EPR spectrometer. Bruker Report 1(1987):22–25
6. The BRUKER FT EPR Spectrometer ESP 380 (1989). In: Keijzers CP, Reijerse EJ, Schmidt J (eds) Pulsed EPR: a new field of applications. Koninklijke Nederlandse Akademie van Wetenschappen, Amsterdam
7. Schmalbein D, Maresch GG, Kamlowski A, Hofer P (1999) The Bruker high-frequency-EPR system. Appl Magn Reson 16(2):185–205. https://doi.org/10.1007/Bf03161933
8. The new technology X-band microwave bridge for CW- and FT-EPR (2001). Bruker report 149/2001, pp 14–15
9. Smith GM, Cruickshank PAS, Bolton DR, Robertson DA (2008) High-field pulse EPR instrumentation. In: Electron paramagnetic resonance, vol 21. The Royal Society of Chemistry, pp 216–233. https://doi.org/10.1039/b807958g
10. Tkach I, Halbmair K, Hobartner C, Bennati M (2014) High-frequency 263 GHz PELDOR. Appl Magn Reson 45(10):969–979. https://doi.org/10.1007/s00723-014-0581-z
11. Poole CP (1983) Electron spin resonance: a comprehensive treatise on experimental techniques/second edition. John Wiley, New York
12. Bender C, Berliner LJ (eds) (2004) EPR: instrumental methods, vol 21. Biological magnetic resonance. Springer US, New York
13. Bender C (2004) The generation and detection of electron spin echoes. In: Bender C, Berliner LJ (eds) EPR: instrumental methods, vol 21. Springer US, New York, pp 212–276. https://doi.org/10.1007/978-1-4419-8951-2
14. Semenov AG, Shchirov MD, Zhidkov VD, Khmelinskii VE, Dvornikov EV (1980) A coherent electron spin echo spectrometer. Preprint No. 3 of the Institute of Chem. Kinetics and Combustion, Novosibirsk
15. Hoefer P, Carl PJ (2006) X-band pulse-EPR resonator performance. Bruker Report 157 (158):52–56
16. Piasecki W, Froncisz W, Hyde JS (1996) Bimodal loop-gap resonator. Rev Sci Instrum 67 (5):1896–1904
17. Rinard GA, Eaton GR (2005) Loop-Gap resonators. In: Eaton SR, Eaton GR, Berliner LJ (eds) Biomedical EPR, Part B: methodology, instrumentation, and dynamics. Springer US, Boston, MA, pp 19–52. https://doi.org/10.1007/0-306-48533-8_2
18. Raitsimring A, Astashkin A, Enemark JH, Blank A, Twig Y, Song Y, Meade TJ (2012) Dielectric resonator for K_a-band pulsed EPR measurements at cryogenic temperatures: probehead construction and applications. Appl Magn Reson 42(4):441–452. https://doi.org/10.1007/s00723-012-0313-1
19. Hyde JS, Chien JCW, Freed JH (1968) Electron-electron double resonance of free radicals in solution. J Chem Phys 48(9):4211–4226. https://doi.org/10.1063/1.1669760
20. Milov AD, Salikhov KM, Shirov MD (1981) Application of the double resonance method to electron spin echo in a study of the spatial distribution of paramagnetic centers in solids. Sov Phys Solid State 23(4):565–569
21. Milov AD, Maryasov AG, Tsvetkov YD (1998) Pulsed electron double resonance (PELDOR) and its applications in free-radicals research. Appl Magn Reson 15(1):107–143. https://doi.org/10.1007/Bf03161886
22. Tkach I, Sicoli G, Hobartner C, Bennati M (2011) A dual-mode microwave resonator for double electron-electron spin resonance spectroscopy at W-band microwave frequencies. J Magn Reson 209(2):341–346
23. Bowman MK (1990) Fourier Transform Electron Spin Resonance. In: Kevan L, Bowman MK (eds) Modern pulsed and continuous electron spin resonance, vol 1st. Wiley, New York, pp 1–42
24. Mizuta Y, Kohno M, Fujii K (1993) Development of microwave control in pulsed-ESR apparatus. Jpn J Appl Phys 1 32(3a):1262–1267. https://doi.org/10.1143/jjap.32.1262
25. Hoefer P, Kamlowski A (2001) The new technology CW/FT-EPR microwave bridge. pulsed electron-electron double resonance (ELDOR) experiments. Bruker report 149/2001, pp 16–18

26. Hoefer P, Heilig R, Maier DC, Prisecaru I, Schmalbein D (2003) The superQ-FT accessory for pulsed EPR, ENDOR and ELDOR at 34 GHz. Bruker Report 152(153):37–43
27. Carl P, Heilig R, Maier DC, Hoefer P, Schmalbein D (2004) The W-band power upgrade module for pulsed EPR, ENDOR and ELDOR at 94 GHz. Bruker Report 154(155):35–40
28. Borbat PP, Crepeau RH, Freed JH (1997) Multifrequency two-dimensional Fourier transform ESR: An X/Ku-band spectrometer. J Magn Reson 127(2):155–167
29. Goldfarb D, Lipkin Y, Potapov A, Gorodetsky Y, Epel B, Raitsimring AM, Radoul M, Kaminker I (2008) HYSCORE and DEER with an upgraded 95 GHz pulse EPR spectrometer. J Magn Reson 194(1):8–15
30. Tsvetkov YD, Grishin YA (2009) Techniques for EPR spectroscopy of pulsed electron double resonance (PELDOR): a review. Instrum Exp Tech. 52(5):615–636. https://doi.org/10.1134/s0020441209050017
31. Prisner TF, Rohrer M, Mobius K (1994) Pulsed 95-Ghz, high-field EPR heterodyne spectrometer with high spectral and time resolution. Appl Magn Reson 7(2–3):167–183. https://doi.org/10.1007/Bf03162610
32. Dubinskii AA, Grishin YA, Savitsky AN, Mobius K (2002) Submicrosecond field-jump device for pulsed high-field ELDOR. Appl Magn Reson 22(3):369–386. https://doi.org/10.1007/Bf03166118
33. Schnegg A, Dubinskii AA, Fuchs MR, Grishin YA, Kirilina EP, Lubitz W, Plato M, Savitsky A, Mobius K (2007) High-field EPR, ENDOR and ELDOR on bacterial photosynthetic reaction centers. Appl Magn Reson 31(1–2):59–98
34. Rohrer M, Brugmann O, Kinzer B, Prisner TF (2001) High-field/high-frequency EPR spectrometer operating in pulsed and continuous-wave mode at 180 GHz. Appl Magn Reson 21(3–4):257–274. https://doi.org/10.1007/Bf03162406
35. Hertel MM, Denysenkov VP, Bennati M, Prisner TF (2005) Pulsed 180-GHz EPR/ENDOR/PELDOR spectroscopy. Magn Reson Chem 43:S248–S255
36. Denysenkov VP, Prisner TF, Stubbe J, Bennati M (2005) High-frequency 180 GHz PELDOR. Appl Magn Reson 29(2):375–384. https://doi.org/10.1007/Bf03167024
37. Denysenkov VP, Prisner TF, Stubbe J, Bennati M (2006) High-field pulsed electron-electron double resonance spectroscopy to determine the orientation of the tyrosyl radicals in ribonucleotide reductase. P Natl Acad Sci USA 103(36):13386–13390. https://doi.org/10.1073/pnas.0605851103
38. Denysenkov VP, Biglino D, Lubitz W, Prisner TF, Bennati M (2008) Structure of the tyrosyl biradical in mouse R2 ribonucleotide reductase from high-field PELDOR. Angew Chem Int Ed 47(7):1224–1227
39. Biospin B (2017) ELEXSYS E780. https://www.bruker.com/fileadmin/user_upload/8-PDF-Docs/MagneticResonance/EPR_brochures/Elexsys_E780_flyer_0714_T13102_lo-res.pdf. Accessed 14 Mar 2017
40. Rengan SK, Bhagat VR, Sastry VSS, Venkataraman B (1969) Magnetic field-pulsed ELDOR spectroscopy. J Magn Reson 33(2):227–240. https://doi.org/10.1016/0022-2364(79)90242-7
41. Dzuba SA, Tsvetkov YD (1982) Slow rotations of nitroxyl radicals in viscous liquids studied by ESE. Khim Fiz 9:1197–1201
42. Dzuba SA, Maryasov AG, Salikhov KM, Tsvetkov YD (1984) Superslow rotations of nitroxide radicals studied by pulse EPR spectroscopy. J Magn Reson 58(1):95–117
43. Kispert LD (2005) Electron-Electron Double Resonance. In: Eaton SR, Eaton GR, Berliner LJ (eds) Biomedical EPR, Part B: methodology, instrumentation, and dynamics. Springer US, Boston, MA, pp 165–197. https://doi.org/10.1007/0-306-48533-8_6
44. Kulik LV, Grishin YA, Dzuba SA, Grigoryev IA, Klyatskaya SV, Vasilevsky SF, Tsvetkov YD (2002) Electron dipole-dipole ESEEM in field-step ELDOR of nitroxide biradicals. J Magn Reson 157(1):61–68

45. Mobius K, Savitsky A, Schnegg A, Plato M, Fuchs M (2005) High-field EPR spectroscopy applied to biological systems: characterization of molecular switches for electron and ion transfer. Phys Chem Chem Phys 7(1):19–42
46. Davis JL, Mims WB (1981) Use of a microwave delay-line to reduce the dead time in electron-spin echo envelope spectroscopy. Rev Sci Instrum 52(1):131–132. https://doi.org/10.1063/1.1136423
47. Narayana PA, Massoth RJ, Kevan L (1982) Active microwave delay-line for reducing the dead time in electron-spin echo spectrometry. Rev Sci Instrum 53(5):624–626. https://doi.org/10.1063/1.1137021
48. Cho H, Pfenninger S, Gemperle C, Schweiger A, Ernst RR (1989) Zero deadtime pulsed ESR by remote echo detection. Chem Phys Lett 160(4):391–395. https://doi.org/10.1016/0009-2614(89)87616-X
49. Doan PE, Hoffman BM (1997) Making hyperfine selection in Mims ENDOR independent of deadtime. Chem Phys Lett 269(3–4):208–214
50. Pfenninger S, Forrer J, Schweiger A, Weiland T (1988) Bridged loop gap resonator—a resonant structure for pulsed electron-spin-resonance transparent to high-frequency radiation. Rev Sci Instrum 59(5):752–760. https://doi.org/10.1063/1.1139822
51. Fauth JM, Schweiger A, Ernst RR (1989) Recovery of broad hyperfine lines in electron spin-echo envelope modulation spectroscopy of disordered-systems. J Magn Reson 81 (2):262–274. https://doi.org/10.1016/0022-2364(89)90058-9
52. Volino F, Csakvary F, Servozga P (1968) Resonant helices and their application to magnetic resonance. Rev Sci Instrum 39(11):1660–1665. https://doi.org/10.1063/1.1683198
53. Mehring M, Freysoldt F (1980) A slotted tube resonator Str for pulsed electron-spin-resonance and odmr experiments. J Phys E: Sci Instrum 13(8):894–895. https://doi.org/10.1088/0022-3735/13/8/022
54. Hyde JS, Froncisz W (1989) Loop gap resonators. In: Advanced EPR. Elsevier, Amsterdam, pp 277–305. https://doi.org/10.1016/b978-0-444-88050-5.50012-4
55. Webb RH (1962) Use of traveling wave helices in ESR and double resonance spectrometers. Rev Sci Instrum 33(7):732–737. https://doi.org/10.1063/1.1717946
56. Biehl R (1986) Sensitivity enhancement in EPR. The dielectric ring TE 011 cavity. Bruker report 1/1986, pp 45–47
57. Ivanov MY, Nadolinny VA, Bagryanskaya EG, Grishin YA, Fedin MV, Veber SL (2016) Bismuth germanate as a perspective material for dielectric resonators in EPR spectroscopy. J Magn Reson 271(Supplement C):83–89. https://doi.org/10.1016/j.jmr.2016.08.009
58. Pfenninger S, Froncisz W, Forrer J, Luglio J, Hyde JS (1995) General-method for adjusting the quality factor of EPR resonators. Rev Sci Instrum 66(10):4857–4865
59. Schweiger A, Jeschke G (2001) Principles of pulse electron paramagnetic resonance. Oxford University Press, Oxford, UK, New York
60. Rinard GA, Quine RW, Eaton SS, Eaton GR (2004) Frequency dependence of EPR sensitivity. In: Berliner LJ, Bender CJ (eds) EPR: instrumental methods. Springer US, Boston, MA, pp 115–154. https://doi.org/10.1007/978-1-4419-8951-2_3
61. Romanelli M, Kurshev V, Kevan L (1994) Comparative-analysis of pulsed electron-spin-resonance spectrometers at X-band and S-band. Appl Magn Reson 7(2–3):427–441. https://doi.org/10.1007/Bf03162623
62. Davoust CE, Doan PE, Hoffman BM (1996) Q-band pulsed electron spin-echo spectrometer and its application to ENDOR and ESEEM. J Magn Reson Ser A 119(1):38–44
63. Salikhov KM, Tsvetkov YD (1979) Electron spin echo studies of spin-spin interactions in solids. In: Kevan L, Schwartz RN (eds) Time domain electron spin resonance. John Wiley, New York, pp 231–278

64. El Mkami H, Ward R, Bowman A, Owen-Hughes T, Norman DG (2014) The spatial effect of protein deuteration on nitroxide spin-label relaxation: Implications for EPR distance measurement. J Magn Reson 248:36–41. https://doi.org/10.1016/j.jmr.2014.09.010
65. Ward R, Bowman A, Sozudogru E, El-Mkami H, Owen-Hughes T, Norman DG (2010) EPR distance measurements in deuterated proteins. J Magn Reson 207(1):164–167. https://doi.org/10.1016/j.jmr.2010.08.002
66. Jeschke G, Bender A, Paulsen H, Zimmermann H, Godt A (2004) Sensitivity enhancement in pulse EPR distance measurements. J Magn Reson 169(1):1–12
67. Ghimire H, McCarrick RM, Budil DE, Lorigan GA (2009) Significantly improved sensitivity of Q-band PELDOR/DEER experiments relative to X-band is observed in measuring the intercoil distance of a leucine zipper motif peptide (GCN4-LZ). Biochemistry-Us 48 (25):5782–5784
68. Zou P, Mchaourab HS (2010) Increased sensitivity and extended range of distance measurements in spin-labeled membrane proteins: Q-band double electron-electron resonance and nanoscale bilayers. Biophys J 98(6):L18–L20. https://doi.org/10.1016/j.bpj.2009.12.4193
69. Swanson MA, Kathirvelu V, Majtan T, Frerman FE, Eaton GR, Eaton SS (2011) Electron transfer flavoprotein domain II orientation monitored using double electron-electron resonance between an enzymatically reduced, native FAD cofactor, and spin labels. Protein Sci 20(3):610–620. https://doi.org/10.1002/pro.595
70. Di Valentin M, Albertini M, Zurlo E, Gobbo M, Carbonera D (2014) Porphyrin triplet state as a potential spin label for nanometer distance measurements by PELDOR spectroscopy. J Am Chem Soc 136(18):6582–6585. https://doi.org/10.1021/ja502615n
71. Kandrashkin YE, van der Est A (2011) Stimulated electron spin polarization in strongly coupled triplet-doublet spin pairs. Appl Magn Reson 40(2):189–204. https://doi.org/10.1007/s00723-011-0194-8
72. Shevelev GY, Krumkacheva OA, Lomzov AA, Kuzhelev AA, Trukhin DV, Rogozhnikova OY, Tormyshev VM, Pyshnyi DV, Fedin MV, Bagryanskaya EG (2015) Triarylmethyl Labels: toward improving the accuracy of EPR nanoscale distance measurements in DNAs. J Phys Chem B 119(43):13641–13648
73. Goldfarb D (2014) Gd^{3+} spin labeling for distance measurements by pulse EPR spectroscopy. Phys Chem Chem Phys 16(21):9685–9699. https://doi.org/10.1039/c3cp53822b
74. Shevelev GY, Lomzov AA, Pyshnyi DV, Kuzhelev AA, Krumkacheva OA, Fedin MV, Rogozhnikova OY, Trukhin DV, Troitskaya TI, Tormyshev VM, Bagryanskaya EG (2015) Spin-labeled oligonucleotides—useful tool for the structural biology. FEBS J 282:350
75. Shevelev GY, Krumkacheva OA, Lomzov AA, Kuzhelev AA, Rogozhnikova OY, Trukhin DV, Troitskaya TI, Tormyshev VM, Fedin MV, Pyshnyi DV, Bagryanskaya EG (2014) Physiological-temperature distance measurement in nucleic acid using triarylmethyl-based spin labels and pulsed dipolar EPR spectroscopy. J Am Chem Soc 136 (28):9874–9877
76. Bagryanskaya EG, Krumkacheva OA, Fedin MV, Marque SRA (2015) Development and application of spin traps, spin probes, and spin labels. Methods Enzymol 563:365–396
77. Kaminker I, Tkach I, Manukovsky N, Huber T, Yagi H, Otting G, Bennati M, Goldfarb D (2013) W-band orientation selective DEER measurements on a Gd^{3+}/nitroxide mixed-labeled protein dimer with a dual mode cavity. J Magn Reson 227:66–71
78. Lovett JE, Lovett BW, Harmer J (2012) DEER-stitch: combining three- and four-pulse DEER measurements for high sensitivity, deadtime free data. J Magn Reson 223:98–106
79. Salvadori E, Fung MW, Hoffmann M, Anderson HL, Kay CW (2015) Exploiting the symmetry of the resonator mode to enhance PELDOR sensitivity. Appl Magn Reson 46 (4):359–368. https://doi.org/10.1007/s00723-014-0621-8

80. Srivastava M, Anderson CL, Freed JH (2016) A new wavelet denoising method for selecting decomposition levels and noise thresholds. IEEE Access 4:3862–3877. https://doi.org/10.1109/ACCESS.2016.2587581
81. Srivastava M, Georgieva ER, Freed JH (2017) A new wavelet denoising method for experimental time-domain signals: pulsed dipolar electron spin resonance. J Phys Chem A 121(12):2452–2465. https://doi.org/10.1021/acs.jpca.7b00183
82. Daubechies I (1992) Ten lectures on wavelets. CBMS-NSF regional conference series in applied mathematics. Society for industrial and applied mathematics. https://doi.org/10.1137/1.9781611970104
83. Mailer C, Haas DA, Hustedt EJ, Gladden JG, Robinson BH (1991) Low-power electron-paramagnetic resonance spin-echo spectroscopy. J Magn Reson 91(3):475–496. https://doi.org/10.1016/0022-2364(91)90375-4
84. Blok H, Akimoto I, Milikisyants S, Gast P, Groenen EJJ, Schmidt J (2009) FID detection of EPR and ENDOR spectra at high microwave frequencies. J Magn Reson 201(1):57–60. https://doi.org/10.1016/j.jmr.2009.08.002

Chapter 3
Nitroxyl Biradicals

Abstract Nitroxyl radicals, also known as nitroxide, imidoxyl and aminoxyl radicals, are ubiquitous in PELDOR and DEER studies because of their stability, solubility, and ability to be chemically modified. They provided the molecular testbed for the measurement of distance and distance distributions; the measurement of weak exchange interactions; and the determination of the mutual orientation of radical fragments. As soon as reliable synthetic methods were developed for stable nitroxyl radicals [1], these free radicals, also known as nitroxide, imidoxyl and aminoxyl radicals, became widely used in chemical and physical studies employing CW EPR spectroscopy. This work is described in detail in monographs [2–7], numerous reviews, and collected volumes, e.g., [8, 9]. The development of pulsed EPR, including ESE, spectroscopy and subsequent applications to physical, chemical, and biological problems also made extensive use of nitroxyl radicals [10, 11]. Nitroxyl radicals and the paramagnetic >N-O· moiety are ubiquitous in PELDOR studies over the last 30 years because of their chemical and biological stability; their solubility in polar and non-polar media; their ability to be chemically modified; and the strong localization of the unpaired electron to the N-O· bond. Nitroxyl biradicals provided the first demonstration of PELDOR modulation from dipole interactions [12]. They provided the molecular testbed for the development of methods to measure the distance between spins and the distance distribution; to measure exchange interactions; and to determine the mutual orientation of radical fragments and other structural features. Today nitroxyl radicals continue to play a major role as spin labels in many PELDOR applications and their EPR properties influence the evolution of PELDOR methods. This Chapter considers some of the nitroxyl radicals that are prominent in PELDOR.

3.1 Nitroxyl Bi-, Tri- and Tetra-radicals

Nitroxyl biradicals furnished the model systems on which most of the theoretical principles of the PELDOR method were experimentally verified. The most commonly-used biradicals have linear molecular linkers terminated by two

© Springer Nature Switzerland AG 2019
Y. D. Tsvetkov et al., *Pulsed Electron–Electron Double Resonance*,
https://doi.org/10.1007/978-3-030-05372-7_3

piperidine, pyrrolidine, pyrroline, or 3-imidazoline radicals. The structure and properties of such biradicals are extensively studied by CW EPR and are detailed in several monographs [3, 5, 6]. Structures of some of the nitroxyl biradicals used to establish the measurement of distances between >N-O· fragments are given in Table 3.1.

3.1.1 Origins of PELDOR

The PELDOR modulation due to dipolar interactions was first detected in frozen glassy solutions of nitroxyl biradical **3-1** [12]. The distance between unpaired electrons was found either from direct measurement of the modulation period, Fig. 3.1, or from the Pake pattern in its Fourier spectrum. The first PELDOR work with magnetic field jumps also used this biradical [13].

Biradicals **3-1** with R = H and **3-2** helped explore how the duration of pulses in PELDOR affects the distance distribution $F(r)$ Sect. 1.6 [14]. This work improved the reliability of $F(r)$ and the determination of the magnitude and sign of the exchange integral J for biradicals with less than 2 nm between >N-O· fragments. Two conformations of **3-2** were found with similar distances of 1.62 and 1.70 nm between >N-O· groups.

A distance of 1.97 $\pm$ 0.01 nm between the >NO· groups of **3-3** was determined from 3pPELDOR Fourier spectra [15]. The same value was also found for the closely-related biradical **3-4** by 3pPELDOR [16] and 4pPELDOR [17], indicating virtually identical dimensions and conformations for **3-3** and **3-4**. In fact, these two biradicals are often confused with each other in the literature, but Gunnar Jeschke indicated to us that they are indeed different. Different groups consistently measure 1.97 nm between the >N-O· groups of **3-4** and report quite narrow lines, ~3–4 MHz, in the PELDOR Fourier spectrum, which agrees with molecular modelling [16] and indicates a single conformation for **3-4** in a glassy matrix at 80 K. The first test of 4pPELDOR was on **3-4** [17].

Nitroxyl biradicals were the molecular ruler to compare distances measured by PELDOR with those known from molecular structures, and to determine the experimental distance limits of PELDOR [16, 18, 21, 23, 24]. The bridging groups connecting the radical fragments provided precise control of distances. Biradicals **3-5** with aliphatic bridges were studied at 80 K in a polystyrene matrix [16]. Although peaks in the Fourier spectra are fairly broad, Fig. 3.2a, their maxima progressively shift to lower frequencies as the chain length increases. Distances from 2.04 nm at $n = 8$ to 3.28 nm at $n = 20$ are obtained [16]. Each link in the chain increases the distance between >NO· fragments in **3-5** by ~0.1 nm. The linewidths in the Fourier spectrum in these experiments result from the set of conformations of the aliphatic chain. The distances determined by PELDOR agree with statistical calculations assuming a completely stretched conformation of the aliphatic chain, Fig. 3.2b.

Table 3.1 Some nitroxyl biradicals studied by PELDOR

	Structure	r, nm *J*, MHz	References
3-1	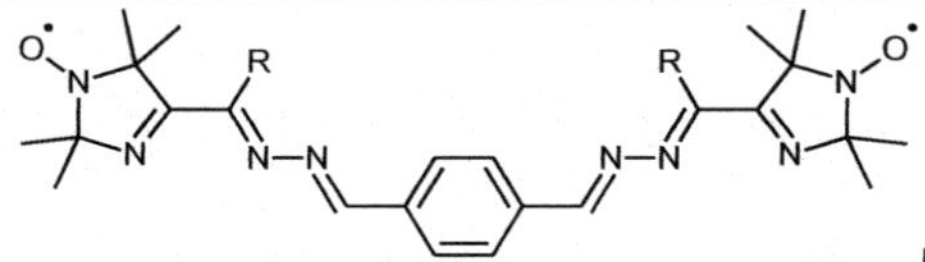 R = H, CH_3	1.9 ± 0.16 1.94 ± 0.02 1.99 ± 0.06 *J* =+0.7 ± 0.4	[12] [14] [14] [14]
3-2	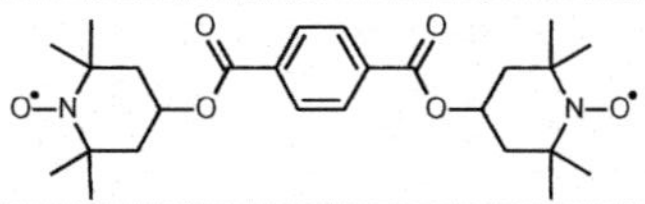	1.62 ± 0.04 1.70 ± 0.04 J=1.4 ± 0.5	[14]
3-3	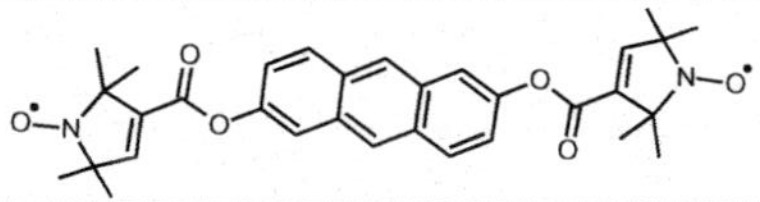	1.97 ± 0.01	[15]
3-4	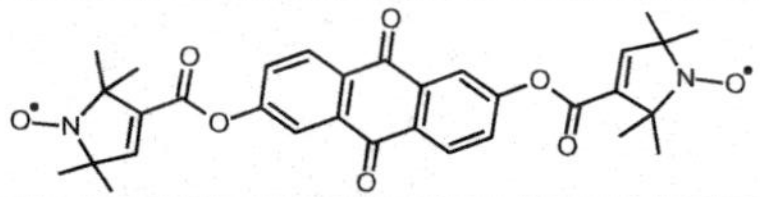	1.98 ± 0.07 1.94 ± 0.05 1.97 ± 0.03	[16] [17] [18]
3-5	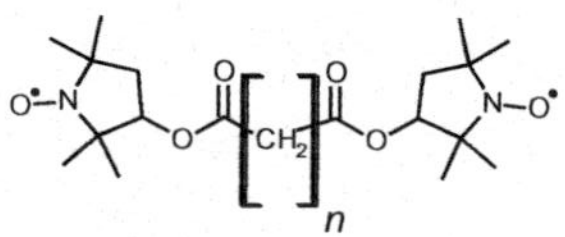 $n = 8$ $n = 12$ $n = 14$ $n = 20$	2.04 ± 0.07 2.48 ± 0.13 2.71 ± 0.19 3.28 ± 0.44	[16]
3-6	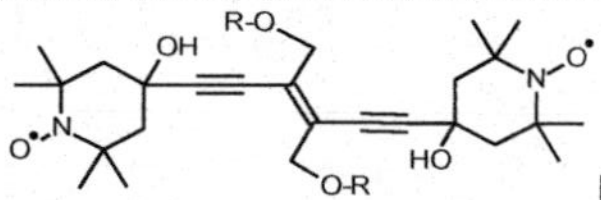 R = *tert*-butyldimethylsilyl	1.54 ± 0.02	[19]

(continued)

Table 3.1 (continued)

	Structure	r, nm *J*, MHz	References
3-7	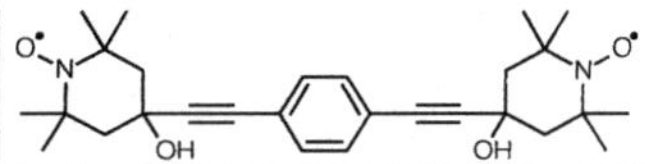	1.72 ± 0.03	[19]
3-8	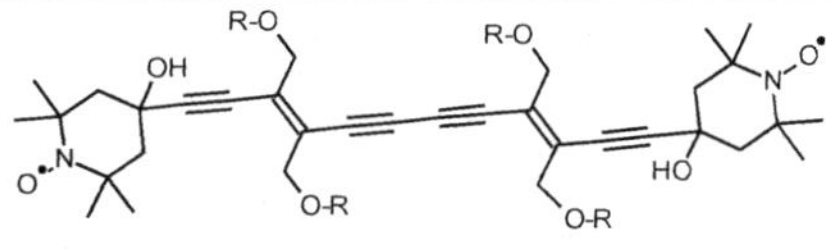 R = *tert*-butyldimethylsilyl	2.20 ± 0.07	[19]
3-9	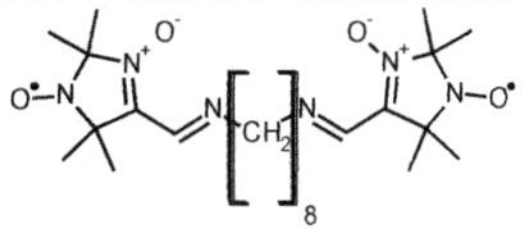	*	[20]
3-10	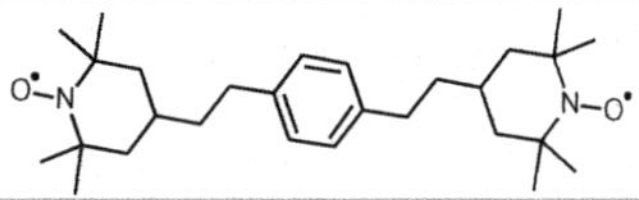	*	[20]
3-11		2.57 ± 0.14	[19]
3-12	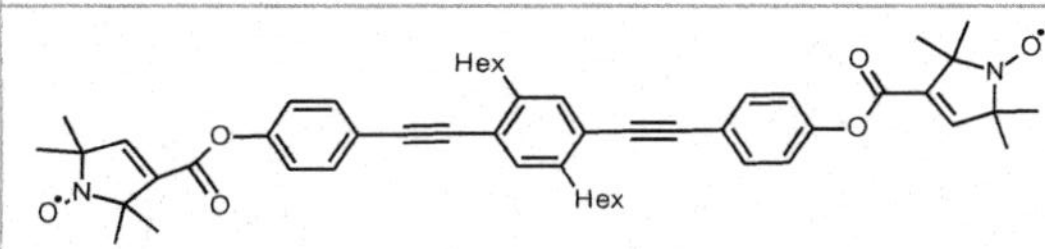 Hex=hexyl	2.83 ± 0.05 2.92 ± 0.04 2.90 ± 0.02	[21] [18] [22]

(continued)

Table 3.1 (continued)

	Structure	r, nm J, MHz	References
3-13	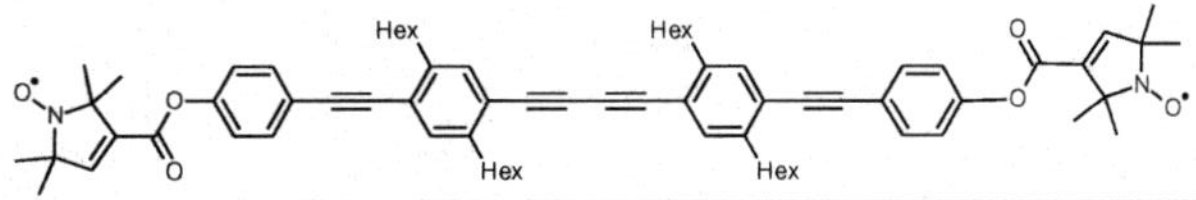	3.63 ± 0.10	[21]
3-14	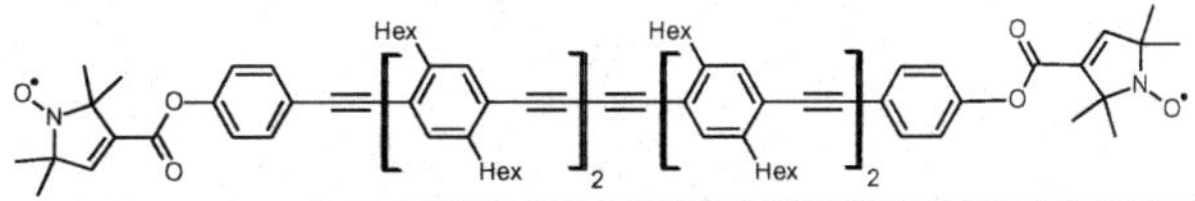	4.92 ± 0.20 5.00 ± 0.40	[21] [23]
3-15	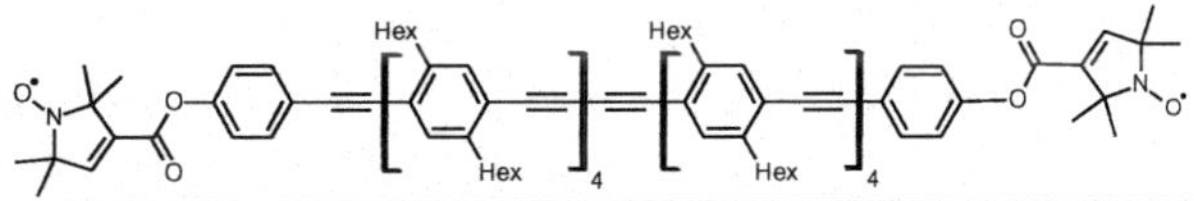	7.44 ± 0.82	[23]
3-16	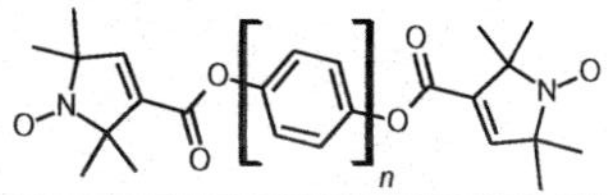 $n = 1$ $n = 2$	1.54 ± 0.03 1.97 ± 0.03	[18]
3-17	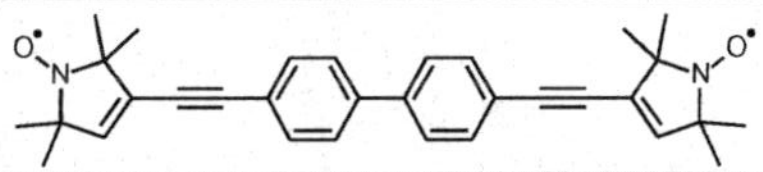	2.40 ± 0.04 J=11.0 ± 0.3	[18]

* No modulation was seen and no distance was obtained

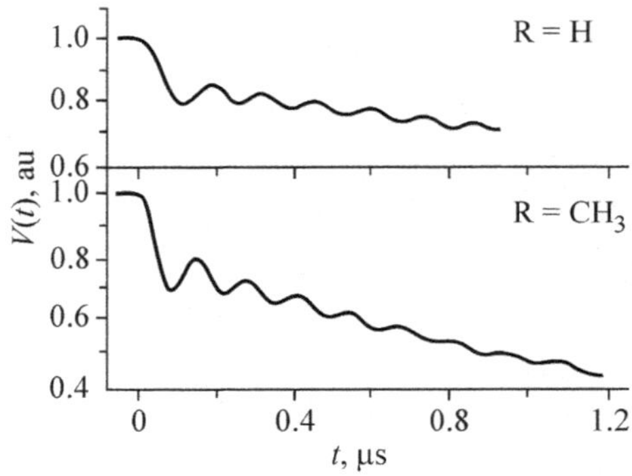

Fig. 3.1 PELDOR time trace of frozen solutions of biradicals **3-1** in toluene Reprinted from Milov et al. [12]

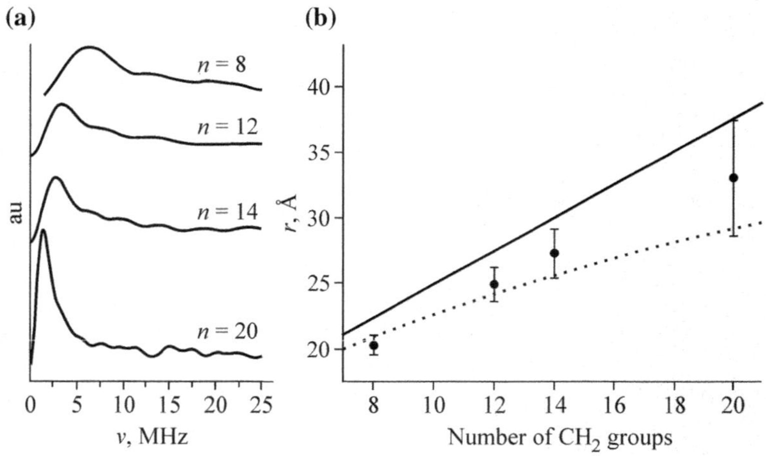

Fig. 3.2 **a** PELDOR Fourier spectra of biradicals **3-5**; **b** distances r between >N-O· groups depend on the number of CH_2 groups in **3-5**: points denote the experiment for n = 8, 12, 14, and 20; the dotted curve shows a statistical calculation of conformations; the solid line shows a completely stretched trans-structure. Reprinted from Pfannebecker et al. [16]

The PELDOR spectra of **3-5** have unusually good resolution. Biradicals **3-9** and **3-10**, with similar molecular structures, lack modulation beats in their 3pPELDOR time traces and the Fourier spectra lack distinct peaks [20]. In our opinion, the aliphatic bridge in these frozen biradical solutions adopt a more diverse set of conformations than for **3-5** in polystyrene. Subsequent PELDOR studies focus on biradicals with rigid, conformationally-fixed bridging groups.

3.1.2 The Outer Limits

The upper boundary of distances that can be measured by PELDOR using nitroxyl radicals was explored with biradicals **3-11** to **3-15** having a rigid molecular bridge composed of triple bonds and benzene rings [21, 23]. The largest distance of 7.44 ± 0.82 nm was measured for **3-15** in deuterated *ortho*-terphenyl at 50 K [23]. Good accuracy at a distance of 5.0 nm was achieved with **3-14** by PELDOR [19, 23] and by another dipolar spectroscopy known as SIFTER. Both methods gave 5.10 ± 0.12 nm [21], while theoretical MD modelling predicted a distance between 5.15 and 5.23 nm for **3-14**. The excellent correlation between MD and PELDOR methods for the distances between >N-O· fragments also holds for other shorter, but more rigid, biradicals: **3-6** to **3-8**, **3-11**, and **3-12** [18, 19, 21, 25].

Spin relaxation seems to limit the maximum distances that can be measured by PELDOR using nitroxyl labels, while the shortest distances are limited more by instrumental parameters.

3.1.3 Small Clusters of Radicals

Nitroxyl radicals specially linked as pairs, triples, and quadruples provided the most rigorous test of PELDOR theory [26]. The goal was to experimentally verify the theoretical basis for determining the number of spins in a group and their pair distribution function for fixed values of $N = 2$, 3, and 4.

The rather rigid **3-18** to **3-22** containing precise numbers of spins were used, Table 3.2. The 4pPELDOR time traces of frozen solutions in deuterated toluene show distinct dipole modulation, Fig. 3.3. The limiting signal amplitude V_p was measured after background subtraction. The pump probability p_B was determined to be 0.43 and 0.12 for pump pulse durations of 12 and 92 ns, respectively, based on PELDOR time traces of **3-16** with $n = 2$. With an accuracy of ~5%, Eq. 1.26 gave experimental values of $N = 2.1$, for biradicals **3-18,** and **3-19**; $N = 3.0$ for triradicals **3-20** and **3-21**, and $N = 3.9$ for tetraradical **3-22**, corresponding completely to the values expected for these polyradicals.

The distance distribution $F(r)$ for polyradicals has some noteworthy features. For biradicals **3-18**, **3-19** and the symmetric triradical **3-20**, $F(r)$ has only one maximum, as expected for structures having the same distance between their nitroxyl groups. However, $F(r)$ for triradical **3-21** contains two lines with an intensity ratio of 2:1, Fig. 3.3c, as expected from the inequivalent positions of nitroxyl groups. For tetraradical **3-22**, $F(r)$ has three maxima, Fig. 3.3d, corresponding to the three possible distances for a rectangle with nitroxyl groups at the vertices. The distances between spins in these polyradicals agree well with MD results given in parentheses, Table 3.2.

A triradical can have up to three dipolar frequencies corresponding to the distances between pairs of spins in the triradical. The PELDOR time trace or its

Table 3.2 Bi-, tri-, and tetra-nitroxyls studied by the PELDOR method in [26, 27]

N	Structure	r, nm	References
3-18	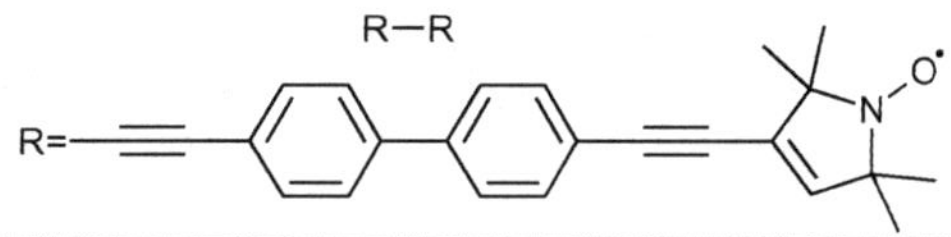	3.4 (3.4)[a]	[26]
3-19	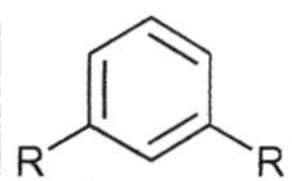	3.3 (3.4)	[26]
3-20	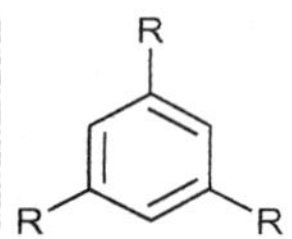	3.3 (3.4)	[26]
3-21	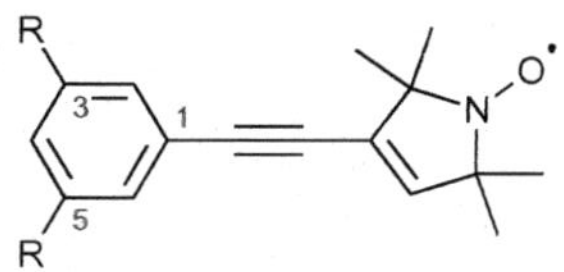	1,3/1,5[b] 2.5 (2.5) 3,5 3.3 (3.4)	[26]
3-22	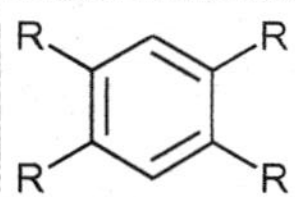	*ortho*[c] 2.2 (2.0) *meta* 3.3 (3.4) *para* 3.8 (3.9)	[26]

(continued)

Table 3.2 (continued)

N	Structure	r, nm	References
3-23	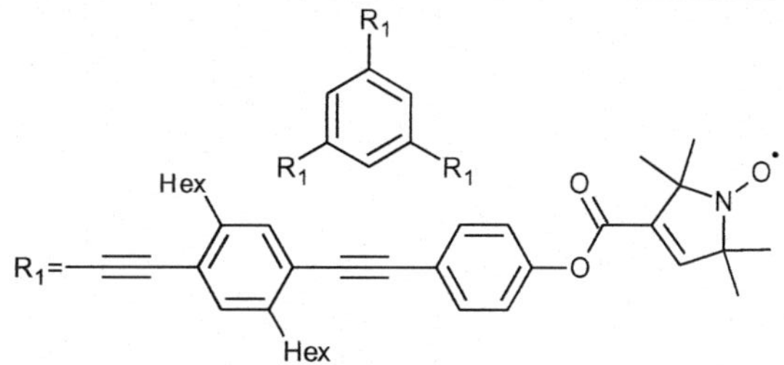	3.52 3.54 3.60	[27]
3-24	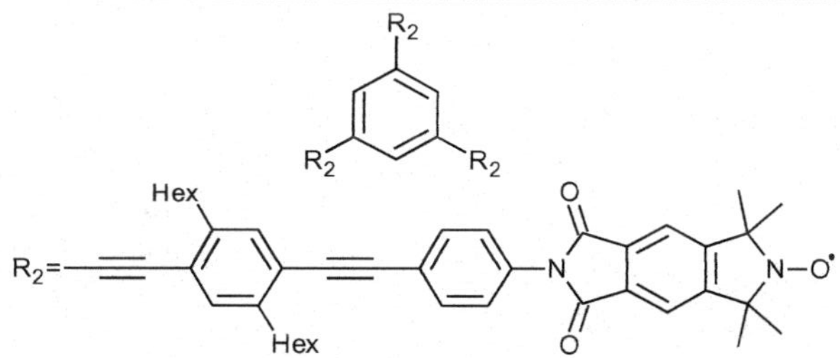	3.83	[27]
3-25	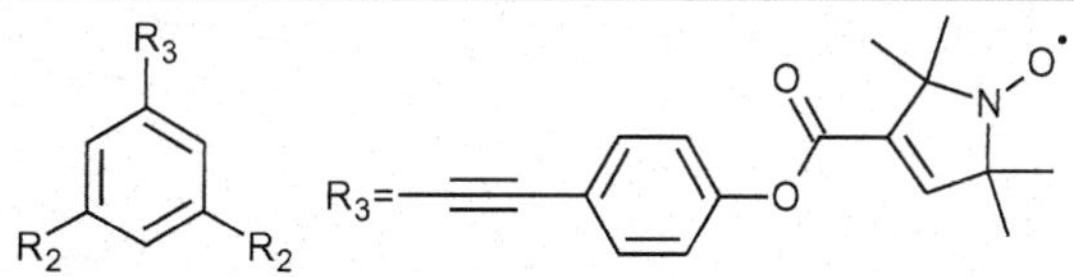	2.90 3.07 3.60	[27]

(continued)

Table 3.2 (continued)

N	Structure	r, nm	References
3-26	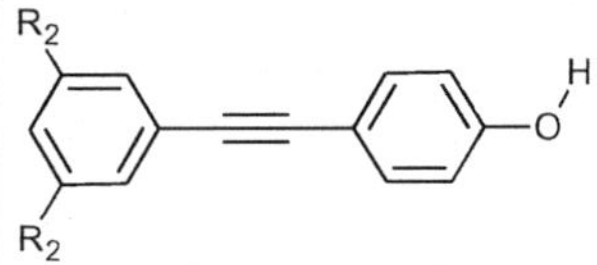	3.83	[27]
3-27	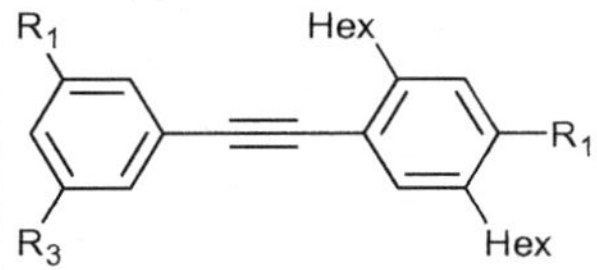	3.23 3.88 4.32	[27]

[a]Calculated distances are given in parentheses
[b]Positions of substituents (radicals) in the benzene ring are given
[c]Distances for radical pairs in *ortho-*, *meta-* and *para-*positions to each other

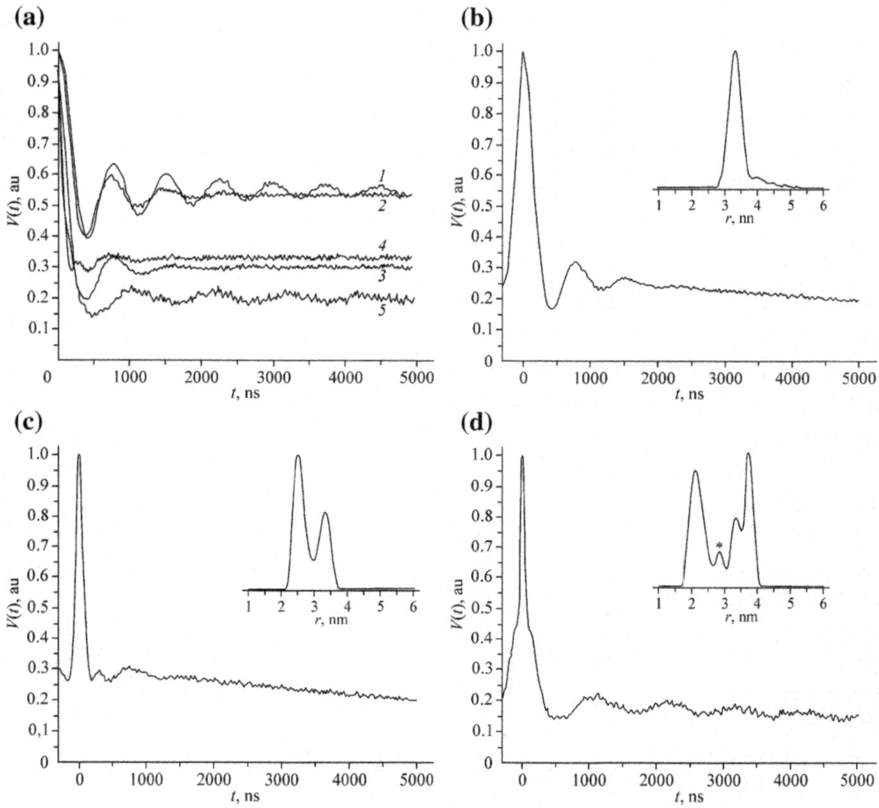

Fig. 3.3 **a** Experimental PELDOR time traces used to determine the number of radicals N in biradicals 1—**3-18**, 2—**3-19**; triradicals 3—**3-20** and 4—**3-21**; and 5—tetraradical **3-22**; **b-d** PELDOR time traces and $F(r)\mathrm{d}r$ for **3-19**, **3-21**, and **3-22**, respectively. Reprinted from Bode et al. [26]

Fourier spectrum contain these dipolar frequencies, but in addition, combinations in the form of sums and differences of these frequencies can be present. If the combination frequencies fall in the bandwidth of the PELDOR measurements, they appear as additional lines in both the Fourier spectrum and $F(r)$, or broaden the line in $F(r)$. It is possible to separate combination from dipolar frequencies in polyradicals from measurements for several values of p_B [27]. Because combination frequencies arise from dipolar frequencies in a single polyradical, they could provide a new way to correlate spins within the same polyradical.

4pPELDOR measurements of bi- and triradicals **3-23** to **3-27**, Table 3.2 [28], were made in glassy solutions of perdeuterated *ortho*-terphenyl at 50 K [27]. The dependence of the modulation on p_B was used to eliminate the combination frequencies. The $F(r)$ for triradicals substantially sharpened, and the PELDOR Fourier spectra became sharper without the combination frequencies. It turns out that even in the chemically-symmetric triradical **3-23**, the distances between the three >N-O·

fragments are not equal because each fragment has several possible conformations [27].

Mixtures of bi- and triradicals provided a good model for small clusters of radicals. The wide range of nitroxyl polyradicals developed over the years played a key part in establishing the basis for reliable determination of the number of spins in studies of complexation and aggregation in complicated biological systems.

3.2 Oligomers and Supramolecules with Nitroxyls

In addition to verifying the theoretical foundation of PELDOR and testing the reliability of PELDOR measurements, nitroxyl radicals enabled PELDOR studies of oligomers and supramolecules. That work focused on the structure and flexibility of the large molecular building blocks used in nanostructures and nanomachines. The real size and flexibility of a large, complicated molecule depend on all of its conformations and the energy of each conformation, which are reflected in the histograms of distances and the distance distribution $F(r)$. PELDOR measurements of $F(r)$ for terminal nitroxyl groups attached to oligomers and supramolecules provide a critical comparison of end-to-end distances and persistence lengths obtained from MD calculations and other physical measurements. The conformations and structures, that are present in room temperature solutions, usually are preserved when the solvent is flash frozen as a glass. This allows the full distribution of conformations to be characterized by PELDOR measurements.

3.2.1 p-Phenylene-Ethynylene Nanowires

Long π-conjugated oligomers, so-called molecular wires, are used in organic conductors, sensors, and photoconversion devices. In these applications, the conformational and molecular flexibility of these oligomers is important for structural integrity and for conductivity. Because PELDOR measures distances in a nanometer range and the distance distribution function, it is one of the few methods to determine molecular rigidity.

An important class of nanowires is the *p*-phenylene + ethynylene (PPE) system of carbon-carbon triple bonds interspersed with phenyl groups. PPE oligomers were used as linkers in biradicals **3-12** to **3-15** used to test MD methods for predicting the length and flexibility of such oligomers [29]. This set of nitroxyl-terminated nanowires was extended to **3-28** to **3-31**, Table 3.3, with additional PPE nanowires of varying length [29, 30].

Total flexibility can be split into contributions from conjugated segments, based on MD calculations [29, 30]. This was tested by measurements made at 50 K in a series of glassy solvents with different glass-transition temperatures T_g. The

Table 3.3 Structure of PPE oligomers, Hex can be either hexyl or 6-methoxyhexyl [29, 30]

	Supramolecular system	n
3-28	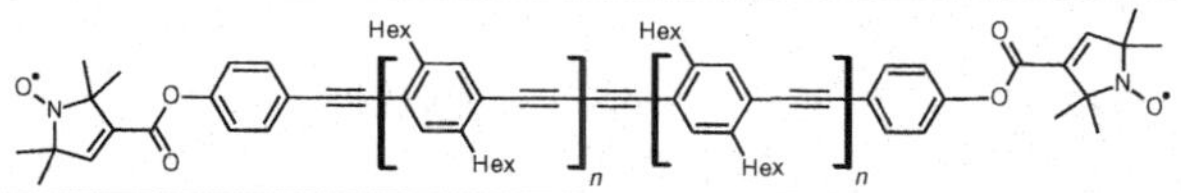	1–4
3-29	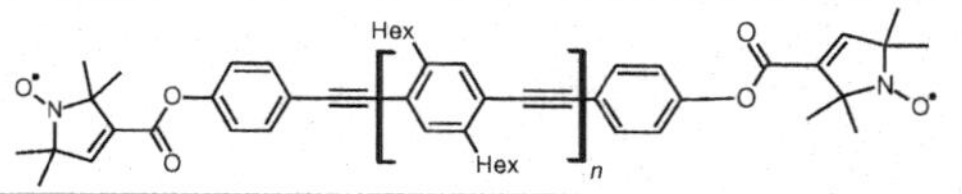	1–5
3-30	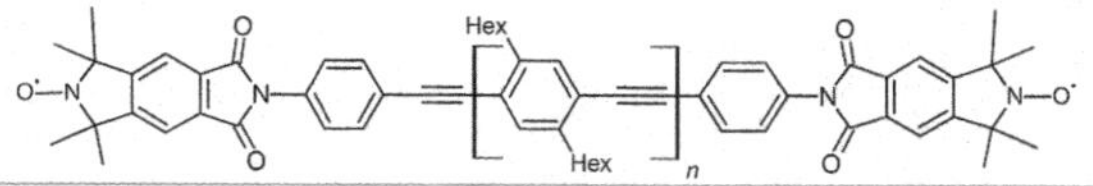	1–4
3-31	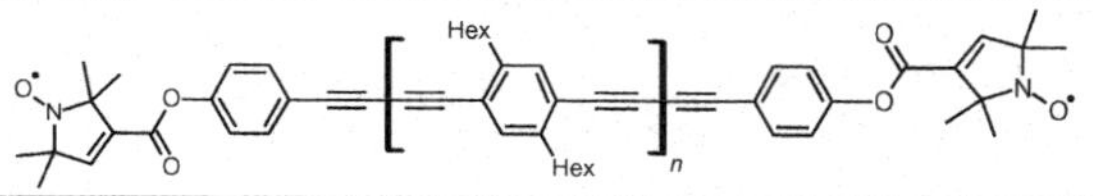	1–4

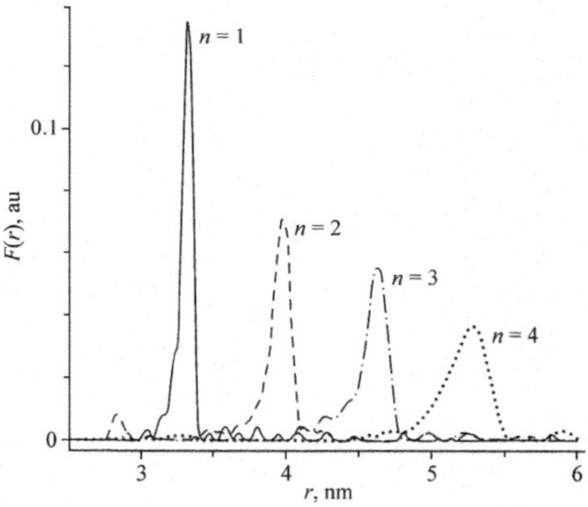

Fig. 3.4 The distance distribution $F(r)$ between >N-O· groups from Tikhonov regularization of PELDOR data for **3-30,** n = 1–4. Reprinted from Godt et al. [28]

harmonic segmented chain (HSC) model predicts that $F(r)$ for linear molecules is asymmetric. The shape and position of peaks in $F(r)$ measured by PELDOR changes systematically with length, Fig. 3.4. and agrees with predictions. The width of the distribution increases substantially as the length increases. All these features are seen for each nanowire type, Fig. 3.4 [29, 30]. The structure of the nitroxyl spin label has only a weak effect, ~10% on the experimental $F(r)$,

3.2.2 *Molecular Ladders*

Another class nanostructures whose flexibility was studied by PELDOR are molecular ladders. They are constructed from helical blocks and are labeled by pyrroline nitroxyl radicals at both ends for PELDOR measurements [31]. Blocks of bis-peptide monomers were formed into ladders **3-32** with n = 4–8, Table 3.4 [31].

Table 3.4 Supramolecular systems studied by the PELDOR method

	Supramolecular system	r, nm	References
3-32	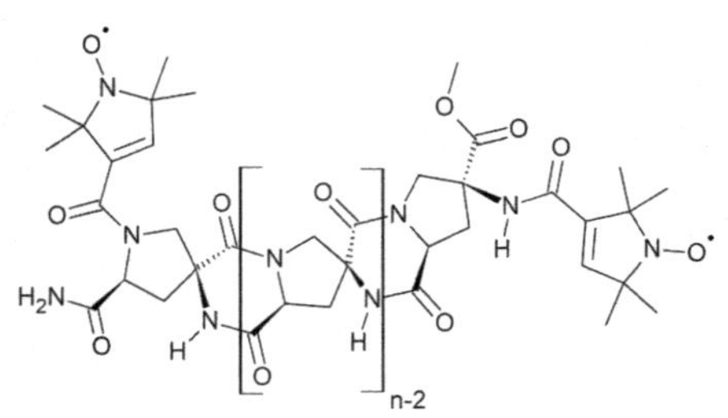 n=4 n=5 n=6 n=7 n=8	2.37 ± 0.10 2.67 ± 0.17 2.94 ± 0.10 3.23 ± 0.14 3.45 ± 0.11	[31, 32]
3-33	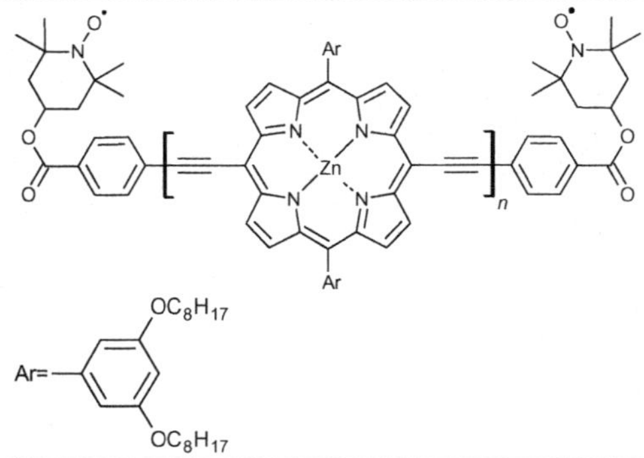 n=1 n=2 n=3 n=4	3.36[a] (3.40) (3.44) 4.64 (4.68) (4.79) 5.88 (5.96) (6.14) 7.34 (7.20) (7.50)	[33]
3-34	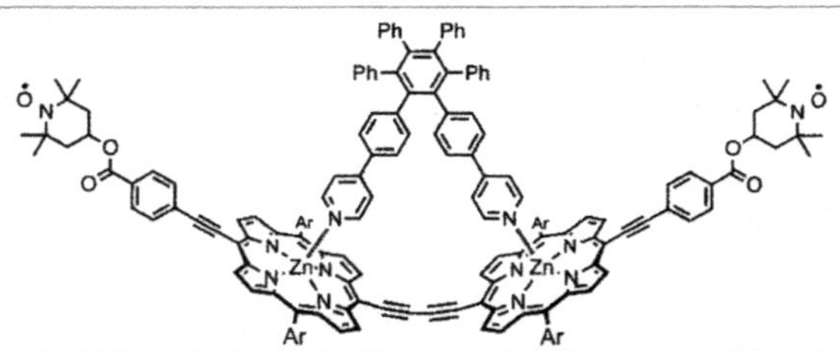	4.15 (4.30)	[33]
3-35	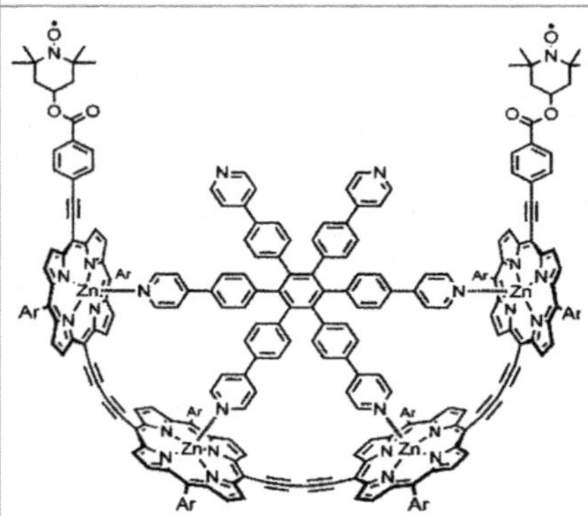	2.49 (2.79)	[33]

(continued)

Table 3.4 (continued)

	Supramolecular system	r, nm	References
3-36	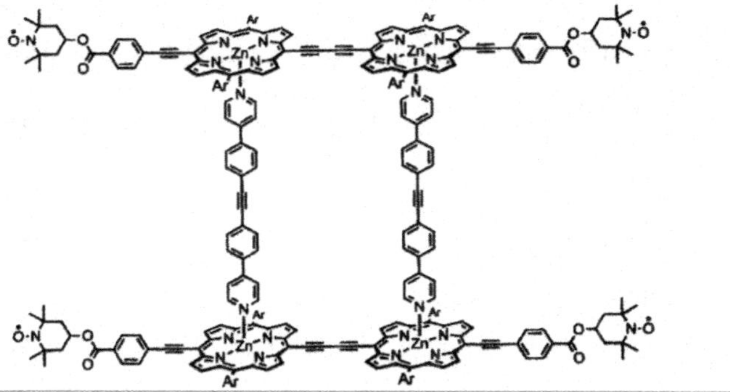	2.69 (2.69) 4.62 (4.70) 4.76 (5.08)	[33]
3-37	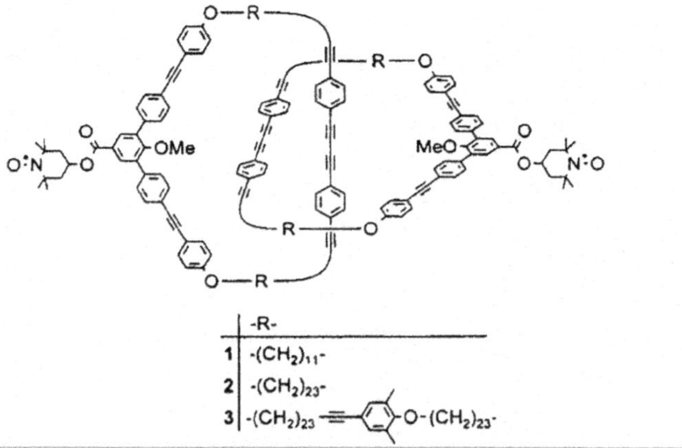	~2.5 ~3.7 ~5.0	[34]

[a]Values from PELDOR, (Values) from MD calculation, *(Values)* from single-crystal X-ray diffraction

These biradical ladders are water-soluble, unlike the biradicals in Table 3.1 which, as a rule, are soluble only in organic solvents. Aqueous solutions of **3-32** were studied at 80 K in buffer with glycerin added for glassing.

The 4pPELDOR Fourier spectra and the $F(r)$ were analysed. As n increases, the average distance between >N-O· groups grows from 2.37 nm at $n = 4$ to 3.46 nm at $n = 8$, with each additional monomer block adding 0.27 nm, on average, to the distance. It is worth noting that the width or standard deviation of the $F(r)$ peak increases with n from 0.18 nm at $n = 4$ to 0.58 nm at $n = 8$. This width characterizes the molecular flexibility of this structure and is consistent with MD calculations.

Bis-peptide ladder structures similar to **3-32**, but with slightly different monomer blocks were synthesized and examined [32]. Depending on the monomers and their placement in the ladder chain, the shape of the ladder could be controlled, from linear to S-shaped or even circular. Both the distance between >N-O· groups and the width of the $F(r)$ peak reflect the conformation spread or flexibility of these ladders. Directed synthesis of such supramolecular ladder structures, especially with monomers having different functional groups, enables the targeted design of useful nanosystems.

PELDOR provides an important, objective method to measure sizes and shapes of these ladders. Theoretical calculations based on a rigid segment model for these bis-peptide ladders agree much better with the $F(r)$ measured by PELDOR than did conventional MD calculations [35].

3.2.3 Porphyrin Oligomers

Porphyrin oligomers can serve as convenient building blocks for supramolecular structures potentially suitable for nanoelectronics and nanorobotics. Nitroxyl radicals are readily attached as spin labels to the ends, Table 3.4, for PELDOR studies of structure and flexibility. Linear supramolecules of butadiyne-linked Zn-porphyrin rings were labeled by a nitroxyl group at each end of **3-33** to **3-36** [33].

These molecules were studied at 50 K in frozen solutions of deuterated toluene and *ortho*-terphenyl with pyridine-d_5 and 4-benzyl pyridine, respectively, added to prevent aggregation. PELDOR time traces were analyzed after averaging to remove orientation selectivity [36], see **3.3**. The $F(r)$ from a series of **3-33** with $n = 1{-}4$ show skewed peaks, Fig. 3.5a, b, like other oligomeric structures and are well described by the worm-like chain (WLC) theory [37], and yield flexibility parameters such as the persistent length. The distances found by PELDOR are in excellent agreement with those obtained from X-ray diffraction crystal structures and from MD simulations, Table 3.4 [37].

Fourier spectra of linear oligomers **3-33** with $n = 1{-}4$ reveal $J = 0$ and the absence of spin exchange. This result was unexpected because porphyrin oligomers are considered to have a highly-delocalized electron system and high polarizability

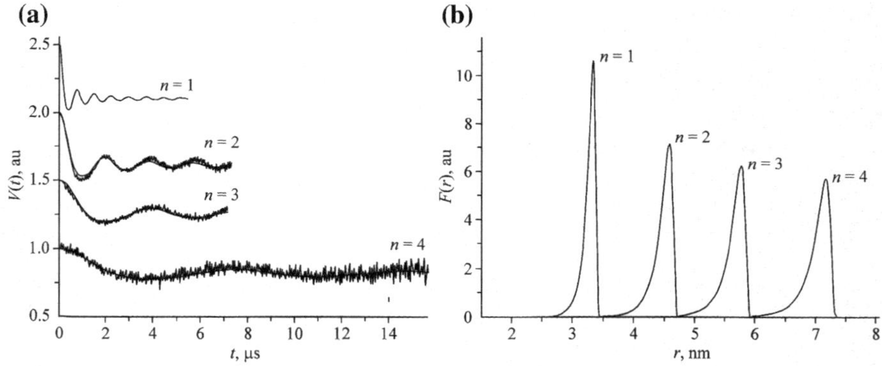

Fig. 3.5 **a** PELDOR time traces for porphyrin chains **3-33** with n = 1–4; **b** the corresponding distance distributions *F(r)*. Reprinted from Lovett et al. [33]

[33], which should produce significant spin exchange. These properties had made porphyrins seem ideal units for the design of molecular conductors.

Some of the properties of supramolecular Zn-porphyrin systems can be tuned by extending the linear geometry into three dimensions by coordination of the Zn, as demonstrated by **3-34** to **3-36** [33]. In **3-34**, two porphyrin centers are coordinated, while four are coordinated in **3-35**. Conformational variations appear, in these complicated, non-linear structures, caused by flexibility of groups containing the nitroxyl group. The conformations produce Gaussian distributions in $F(r)$ rather than the skewed shapes seen in linear systems. For **3-34**, the distance between >N-O· groups is 4.15 nm with a Gaussian width at half maximum of 0.68 nm, while for **3-35**, these values are 2.49 and 0.87 nm. These results agree well with MD calculations.

A more sophisticated situation arises for the four nitroxyl labels in complex **3-36**. Here the geometry of the complex is characterized by three distances: the sides and the diagonal of a rectangle. Distances from PELDOR experiments and MD simulations are given in Table 3.4. The experimental diagonal of 4.76 nm is much less than the 5.08 nm from MD simulations or values for a rectangular structure with the measured sides. This discrepancy is evidence of a twisted conformation for **3-36** in solution. A calculation based on the probable non-planar geometry suggests a twist angle of 44°.

3.2.4 *Inclusion Complexes*

One goal of molecular design is to control the rigidity of elements in molecular machines and mechanisms by understanding the flexibility of supramolecular aggregates. Mobility was studied at 77 K in frozen aqueous glycerol solution for biradical **3-38** and its supramolecular complex with cucurbit [6] uril-based [3] rotaxane **3-39**, Table 3.5 [38]. By itself, **3-38** has a very unconstrained structure, but after it is threaded through a pair of tube-like curcurbit molecules, the conformational flexibility is largely frozen out. This is shown by the lack of modulation in the PELDOR time trace of **3-38** and a very broad distance distribution peaking at $r \geq 2.5$ nm which is supported by MD simulations. In stark contrast, the PELDOR time trace of **3-39** has deep modulation giving $r = 3.07$ nm with a remarkably narrow width of 0.07 nm. The PELDOR method provides experimental estimates of the flexibility of supramolecules based, most importantly, on frozen solutions rather than single crystals [38].

Broad distributions of distances provide important information in some situations, as seen with **3-38** and **3-39**. The [2] catenane supramolecule **3-37**, Table 3.4 [34] provides another example. This supramolecular structure consists of a pair of interlocked rings. Each ring is quite large and is composed of molecular groups that do not have strong interactions with each other, so it is not obvious what its structure might be. The problem of how the rings are oriented relative to each other is easily solved from PELDOR measurements of the $F(r)$ between >N-O· groups.

Table 3.5 Supramolecular inclusion complexes

Supramolecular complex		References
3-38	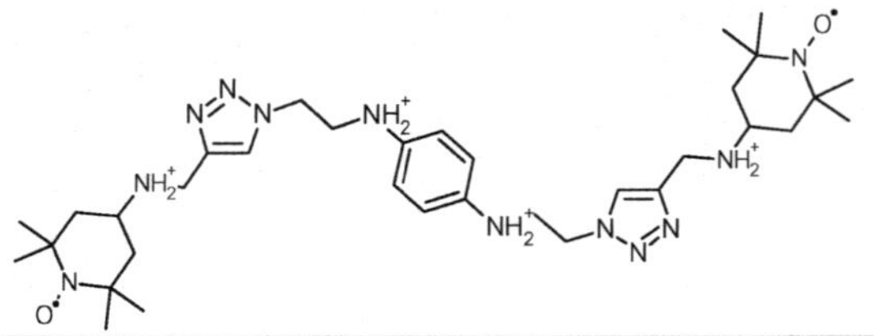	[38]
3-39[a]	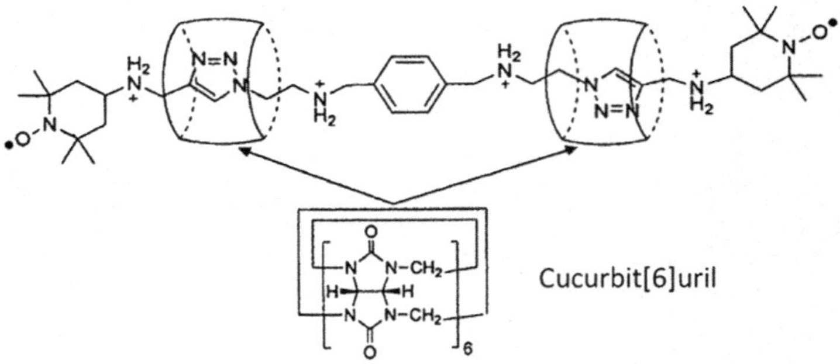	[38]

[a]Reprinted from Pievo et al. [38]

The PELDOR time traces were different at 15 K for frozen solutions of **3-37** in three solvents: chloroform, *o*-terphenyl, and 2-methyltetrahydrofuran. A simulated $F(r)$ was calculated with the >NO groups randomly arranged on two circular rings with an effective radius r_{eff} [34]. Only in chloroform did the experimental data correspond to this simulation. The [2] catenanes aggregated in 2-methyltetrahydrofuran instead of being dispersed as individual supramolecular complexes. There was a substantial distortion of the ring geometry in *o*-terphenyl, likely due to nonlinear conformations of the alkyl part of the rings. The experimental distance distributions in the chloroform solutions, Table 3.4, have widths that reach 4 nm. The predicted r_{eff} values are close to the experimental values seen in chloroform, indicating that the rings do move freely around each other. This simple model describes the properties of molecules several nanometers in diameter and may also be useful even for large supramolecular systems [34].

3.3 Nitroxyl Orientation Selection

Orientation selection in PELDOR means that the PELDOR signal represents some orientations of a radical pair more than others. Orientation selection can provide valuable information on the geometry and dynamics of the pair of spins. At the same time, information about some orientations of the pair of spins may be missing from the PELDOR time trace $V(T)$ when there is orientation selection, information that helps measure distances or $F(r)$. Consequently, standard analysis methods, particularly Tikhonov regularization, can report differing r or $F(r)$ from $V(T)$ measured at different parts of the EPR spectrum or for different values of $\Delta\omega_{AB}$. When orientation selection is not desired, methods have been developed for averaging $V(T)$ obtained for different $\Delta\omega_{AB}$ [29, 39, 40], see Sect. 1.4. This averaging works well with nitroxyl radicals and allows standard PELDOR analysis methods to be used.

3.3.1 Flexible Biradicals

The first studies of orientation selection in PELDOR were performed on **3-3** [15]. If measurements are made by changing the position of the observe or pump pulses in the EPR spectrum, or by changing $\Delta\omega_{AB}$, then the $V(T)$ contain information on the mutual orientation of radicals in the pair [41–44]. These experiments are outlined for a typical nitroxyl EPR spectrum at X-band in Fig. 1.9. Orientation selection for nitroxyl radicals becomes stronger at high frequencies, e.g., Q-, W-bands, where g-factor anisotropy begins to dominate the shape of the EPR spectrum.

The $V(T)$ of biradicals **3-18** and **3-19**, Table 3.2, were measured at X-band for $\Delta\omega_{AB}/2\pi = 40 - 80$ MHz at 40 K in frozen toluene and terphenyl solutions [41]. The modulation frequency and depth does change with $\Delta\omega_{AB}$, as expected when there is orientation selection. A simple model readily accounts for the range of conformations, Fig. 3.6 [41]. Radicals at the ends of the linear **3-18** and bent **3-19** are not oriented randomly, but rotate around the triple bonds so that the N-O· bond lies on a cone with $\alpha = 22°$ for a five-member nitroxyl ring. The linker connecting the radical pairs bends with a random amplitude in another cone with an angle β, characterizing the conformational flexibility of the biradical. The experimental $V(T)$ were best fit with $\beta = 40°$ for **3-18**, while for **3-19**, $\beta = 20°$ in addition to its intrinsic bend of 60° from linearity. PELDOR time traces were calculated numerically and compared to experimental data. The model seems to describe the experimental data well and makes it possible to determine conformational properties of molecules, such as structural flexibility [41].

A more detailed treatment of orientation selection for **3-18** and **3-19** at X-band was attempted [43]. The PELDOR signal was modeled as the convolution of a kernel function containing $F(r)$ with an angular intensity $\lambda(\cos(\theta))$, where θ is the

Fig. 3.6 The geometric model to describe the conformers of biradicals **3-18** and **3-19** [41]

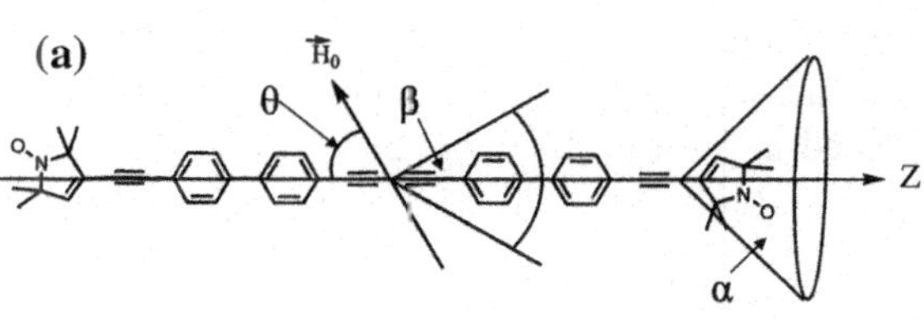

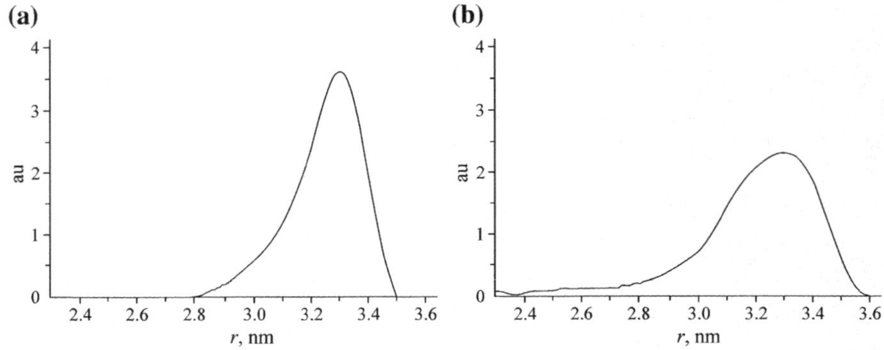

Fig. 3.7 $F(r)$ of nitroxyls **3-18** and **3-19.** Reprinted from Marko et al. [43]

angle between $\vec{r}$ and the vector $\vec{B_0}$ connecting the spins. The function λ is a constant for randomly-oriented pairs without orientation selection. But, with orientation selection, $\lambda(\cos(\theta))$ is the relative contribution from radical pairs at the angle θ. In order to obtain $\lambda(\cos(\theta))$ from experimental data, $V(T)$ must be measured for several $\Delta\omega_{AB}$. Next, it is essential to find $F(r)$ by Tikhonov regularization from a time trace averaged to remove orientation selection. After that, the function $\lambda(\cos(\theta))$ can be found from the measured $V(T)$ for each $\Delta\omega_{AB}$.

The $F(r)$ for **3-18** and **3-19** have a maximum at $\sim$ 3.3 nm, Fig. 3.7. The width of $F(r)$ is much greater for **3-19** than for **3-18** [43] because flexibility changes the distance between spins directly for **3-19**, but only in second order for **3-18**. The asymmetry of $F(r)$ indicates the conformational flexibility in **3-18** and **3-19** [41]. The $\lambda(\cos(\theta))$ lack any distinctive features and it is not clear how to interpret them. This is likely the result of relatively weak orientation selection for nitroxyl radicals at X-band and the spectral overlap of the observe and pump pulses at small $\Delta\omega_{AB}$.

3.3.2 Rigid Biradicals

Orientation selection was used to study the geometry and dynamics of planar, highly-conjugated biradicals **3-40** to **3-42**, Table 3.6 [45, 46]. The nitroxyls are rigidly oriented so that the largest A_{ZZ} component of the hyperfine tensor is perpendicular to the plane of the biradical. The PELDOR time traces and their Fourier spectra varied strongly with $\Delta\omega_{AB}$, an indication of strong orientation selection. The distances between >NO groups were found either from Fourier spectra [45, 46] or by Tikhonov regularization of averaged PELDOR time traces [36]. Examples of orientation selection with other radicals and at higher EPR frequencies are discussed in Sects. 7.2.2 and 8.2.2.

Table 3.6 Planar complexes with strong PELDOR orientation selection [45–47], the z axis of the hyperfine tensor is perpendicular to the plane

Planar complexes	
3-40	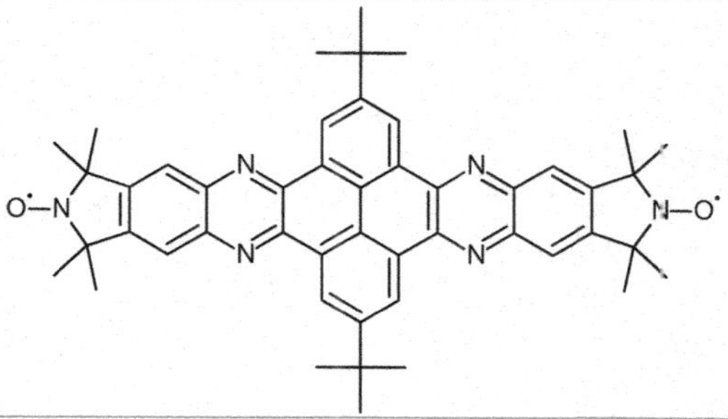
3-41	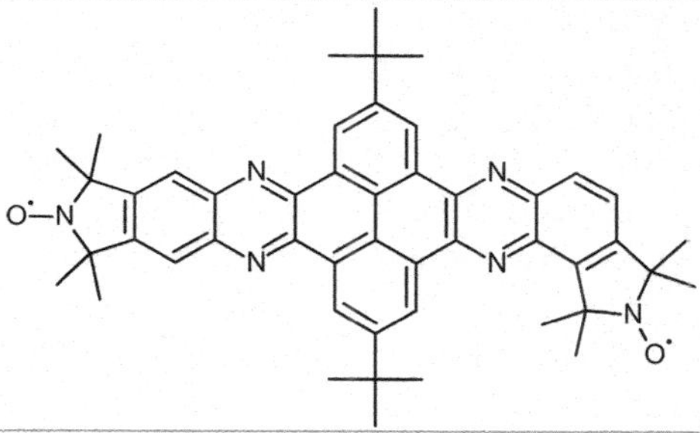
3-42	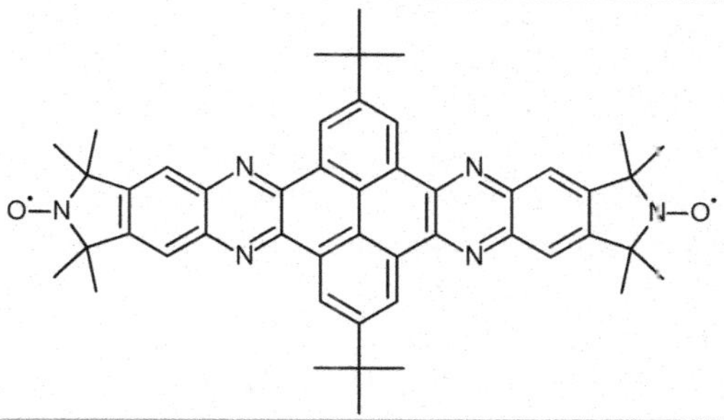

3.4 Exchange Interaction

As a general rule, the exchange interaction is negligible for PELDOR, i.e., $|J| < 1$ MHz, for distances greater than 1.5 nm between unpaired spins. The situation changes if the distance is smaller or if the biradical is part of a rigid system of conjugated bonds and delocalized electrons, in such cases, $J \neq 0$. Moreover, the spin system has antiferromagnetic properties if $J < 0$, but ferromagnetic properties if $J < 0$. Correct determination of the sign of J is important for establishing the magnetic properties of a multi-spin system. The value and sign of the exchange integral can be found from PELDOR Fourier spectra, Fig. 1.4 [24].

A detailed investigation of the value and sign of the exchange integral was performed at X-band on nitroxyl biradicals **3-40** to **3-42** [45, 46]. Their rigid planar structures result in substantial orientation selection. The PELDOR time traces and their Fourier spectra were obtained for different ω_B values. The pump pulse at ω_B was applied at the maximum of the EPR spectrum and the observe pulse at ω_A was applied so that $\Delta\omega_{AB}/2\pi$ ranged from 40 to 80 MHz in 10 MHz steps.

The large A_Z hyperfine component in **3-40** to **3-42** is perpendicular to the biradical plane while the small A_x and A_y lie in the plane. In this situation, measurements at different $\Delta\omega_{AB}$ provide very clear $\omega_\perp$ and $\omega_\|$ features in the PELDOR Fourier spectra, and hence, very accurate r and J Table 3.7. Fourier spectra from the initial PELDOR time traces, as a rule, depend strongly on the sign of J. The sign was verified by comparing calculated Fourier spectra for $+J$ and $-J$ with the experimental spectra. Subsequent W-band PELDOR measurements with even better orientation selection [47] confirmed the X-band results.

The larger distance between nitroxyl groups in **3-40** relative to **3-41** and **3-42** causes the exchange interaction to disappear, $J \approx 0$, while a seemingly trivial structural change from **3-41** to **3-42** reverses the sign of J, changing the spin system from ferromagnetic to antiferromagnetic.

In a detailed study of **3-1**, the value and sign of the exchange integral was estimated to be $J = +0.7 \pm 0.4$ MHz [14]. X- and S-band studies of **3-4**, **3-12**, **3-16**, and **3-17** in frozen toluene-d_8 observed overlapped Fourier spectral lines from dipole or hyperfine interactions [18]. It was difficult to distinguish dipolar frequencies from nuclear modulation in a measurement. However, the nuclear modulation frequencies were identified because only they changed with magnetic field between X and S-band, allowing unambiguous assignment. All biradicals gave frequencies corresponding to $\nu_\perp$ and $\nu_\|$, but only for **3-17** is $\nu_\perp \neq 2\nu_\|$, yielding $J = -11.0$ MHz.

It is interesting that no modulation of the PELDOR time trace was found in either X- or S-bands for a biradical similar to **3-17** but with only one benzene ring in the bridging group [18]. This linker provides a very short, direct conjugation path between the nitroxyl groups, and an exchange integral of $|J| = 73$ MHz is required to simulate the EPR spectrum. The exchange is stronger than the hyperfine interactions (or g-factorm differences), so this biradical is better described in terms of triplet and singlet states rather than weakly interacting doublet states [18]. Differences in the selection rules for EPR transitions for these sets of states will destroy the PELDOR modulation when there is strong exchange.

Table 3.7 PELDOR results for planar biradicals [45–47]

Biradical	r, nm	J, MHz
3-40	2.69 ± 0.1	~0
3-41	1.82 ± 0.4	−3.2 ± 0.6
3-42	1.99 ± 0.4	+2.5 ± 0.5

The link between the two spins has some effect on the magnitude and the sign of the exchange J, even for biradicals with very different spins, e.g., a Cu(II) ion paired with a nitroxyl radical in Sect. 4.2. Methods to analyze PELDOR data from **4-8** [48] were employed to determine the value and sign of J in the closely-related **4-9** [49]. The copper-porphyrin complex **4-9** has a completely conjugated linker to the nitroxyl radical, unlike **4-8** where an ester bond breaks the conjugation. The $V(T)$ from **4-9** at different $\Delta\omega_{AB}$ could be analyzed only by including a weak $J = +4$ MHz [49].

Despite experimental and theoretical difficulties, the PELDOR method will find more and more applications in studies of ion–radical and ion–ion systems that are of particular interest in biophysics and biocatalysis.

3.5 High-Frequency PELDOR

The main advantage for PELDOR spectroscopy in Q-, W-, and higher-frequency bands is the high g-factor resolution. There is a great advantage when only one anisotropic interaction dominates the shape of the spectrum rather than a mixture of g-factor anisotropy and hyperfine interactions from multiple nuclei. Very selective orientation selection measurements become possible simply by changing the position of the observe and pump pulses relative to the g_x, g_y, and g_z features of the EPR spectrum, Fig. 3.8. Higher frequency PELDOR offers greater opportunities to determine the spatial orientation of nitroxyl groups than at X-band.

Methods to measure dipolar interactions by W-band PELDOR were illustrated with **3-1** with R = H and **3-12** containing ^{14}N and ^{15}N [22]. The g_z axes of both >N-O· fragments in **3-1** are parallel to each other and perpendicular to the vector $\vec{r}$ between spins, which are 2.90 ± 0.02 nm apart. Thus, the two nitroxyl rings are coplanar and $\vec{r}$ makes a 26 ± 2° angle with the nitroxyl x axis lying along the N–O bond. The same spatial parameters were found for **3-12**, except the angle between the x axis and $\vec{r}$ is 44°. The authors noted that it is not possible to distinguish *cis*- from *trans*-orientations of the nitroxyl rings for either biradical.

MW power is usually limited at high frequencies, resulting in excitation that is rather selective, with a narrow bandwidth relative to the EPR spectrum, and is well suited for orientation selection measurements, Sect. 3.3. Unfortunately, the low power causes only a slight inversion of the spin magnetization which limits sensitivity and the modulation depth of the PELDOR time trace. These limitations were removed, to a large extent, by using a W-band source with ~1 kW power at 94 GHz and up to 14 ns pulse duration [47].

Six high-power, W-band PELDOR time traces were measured for **3-16** with $n =$ 2, **3-41**, and **3-42**; observing at each g_x, g_y, and g_z spectral line while pumping at each of the other two principal g-values. Experimental $V(T)$ were simulated using spin Hamiltonian parameters from W-band CW EPR spectra and the flexibility angle β, Fig. 3.6. The better resolution of g_x, g_y, and g_z at W-band produced better

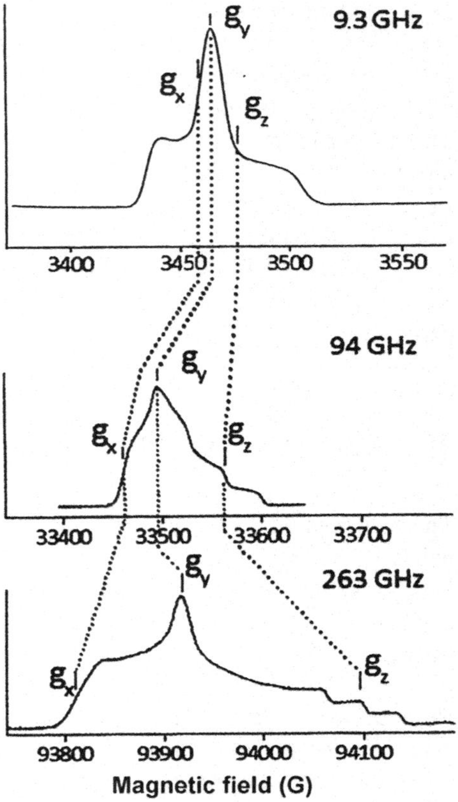

Fig. 3.8 Typical EPR absorption spectrum of a nitroxyl radical at X-, W-, and mm- bands; notice the different magnetic field scales

information on orientation and flexibility in these biradicals while the high power produced more intense modulation and better sensitivity [47].

Very impressive results were achieved at Q-band with 150 W and pulse durations up to 80 ns, with more than an order of magnitude gain in sensitivity compare X-band [40]. Such high power can produce broadband excitation to suppress orientation selection in most cases with nitroxyl radicals, making it easier to determine $F(r)$. On the other hand, orientation selection measurements are still possible. Nuclear modulation from matrix protons and deuterons is strongly suppressed at Q-band, which simplifies analysis of the Q-band time traces and improves signal intensity [40].

References

1. Rozantsev EG (ed) (1970) Free nitroxyl radicals. 1 edn. Springer US, New York. https://doi.org/10.1007/978-1-4757-0710-6
2. Buchachenko AL (1963) Stable radicals. Izdatelstvo Akademii Nauk SSSR, Moskva

3. Rodionov VA, Rozantsev EG (1972) Long-Lived free radicals. Nauka, Moscow
4. Lebedev YS, Muromtsev VI (1970) EPR and relaxation of stabilized radicals (EPR and relaxation of stabilized radicals). Khimiia, Moscow
5. Buchachenko AL, Vasserman AN (1973) Stable radicals. Khimiia, Moscow
6. Parmon VN, Kokorin AI, Zhidomirov GM (1980) Stable biradicals. Nauka, Moscow
7. Vasserman AN, Kovarskii AL (1986) Spin probes and labels in physical chemistry of polymers. Nauka, Moscow
8. Kokorin AI (ed) (2012) Nitroxides - theory. InTech, Moscow, Experiment and Applications. https://doi.org/10.5772/2887
9. Volodarsky LB, Reznikov VA, Ovcharenko VI (1993) Synthetic chemistry of stable nitroxides. CRC Press, Boca Raton
10. Salikhov KM, Semenov AG, Tsvetkov YD (1976) Electron spin echo and its applications. Nauka, Novosibirsk
11. Dikanov SA, Tsvetkov Y (1992) Electron spin echo envelope modulation (ESEEM) spectroscopy. CRC Press, Boca Raton
12. Milov AD, Ponomarev AB, Tsvetkov YD (1984) Modulation beats of signal of double electron-electron resonance in spin-echo for biradical systems. J Struct Chem 25(5):710–713. https://doi.org/10.1007/Bf00747913
13. Kulik LV, Grishin YA, Dzuba SA, Grigoryev IA, Klyatskaya SV, Vasilevsky SF, Tsvetkov YD (2002) Electron dipole-dipole ESEEM in field-step ELDOR of nitroxide biradicals. J Magn Reson 157(1):61–68
14. Milov AD, Naumov BD, Tsvetkov YD (2004) The effect of microwave pulse duration on the distance distribution function between spin labels obtained by PELDOR data analysis. Appl Magn Reson 26(4):587–599. https://doi.org/10.1007/Bf03166585
15. Larsen RG, Singel DJ (1993) Double Electron-Electron Resonance Spin-Echo Modulation: spectroscopic measurement of electron-spin pair separations in orientationally disordered solids. J Chem Phys 98(7):5134–5146. https://doi.org/10.1063/1.464916
16. Pfannebecker V, Klos H, Hubrich M, Volkmer T, Heuer A, Wiesner U, Spiess HW (1996) Determination of end-to-end distances in oligomers by pulsed EPR. J Phys Chem-Us 100 (32):13428–13432. https://doi.org/10.1021/jp960895v
17. Pannier M, Veit S, Godt A, Jeschke G, Spiess HW (2000) Dead-time free measurement of dipole-dipole interactions between electron spins. J Magn Reson 142(2):331–340. https://doi.org/10.1006/jmre.1999.1944
18. Weber A, Schiemann O, Bode B, Prisner TF (2002) PELDOR at S- and X-band frequencies and the separation of exchange coupling from dipolar coupling. J Magn Reson 157(2):277–285. https://doi.org/10.1006/jmre.2002.2596
19. Martin RE, Pannier M, Diederich F, Gramlich V, Hubrich M, Spiess HW (1998) Determination of end-to-end distances in a series of TEMPO diradicals of up to 2.8 nm length with a new four-pulse double electron electron resonance experiment. Angew Chem Int Ed Engl 37 (20):2833–2837. https://doi.org/10.1002/(sici)1521-3773(19981102)37:20%3c2833::aid-anie2833%3e3.0.co;2-7
20. Milov AD, Ponomarev AB, Tsvetkov YD (1984) Electron electron double-resonance in electron-spin echo—model biradical systems and the sensitized photolysis of decalin. Chem Phys Lett 110(1):67–72. https://doi.org/10.1016/0009-2614(84)80148-7
21. Jeschke G, Pannier M, Godt A, Spiess HW (2000) Dipolar spectroscopy and spin alignment in electron paramagnetic resonance. Chem Phys Lett 331(2–4):243–252
22. Savitsky A, Dubinskii AA, Zimmermann H, Lubitz W, Mobius K (2011) High-field dipolar electron paramagnetic resonance (EPR) spectroscopy of nitroxide biradicals for determining three-dimensional structures of biomacromolecules in disordered solids. J Phys Chem B 115 (41):11950–11963. https://doi.org/10.1021/jp206841v
23. Jeschke G, Bender A, Paulsen H, Zimmermann H, Godt A (2004) Sensitivity enhancement in pulse EPR distance measurements. J Magn Reson 169(1):1–12

24. Milov AD, Maryasov AG, Tsvetkov YD (1998) Pulsed electron double resonance (PELDOR) and its applications in free-radicals research. Appl Magn Reson 15(1):107–143. https://doi.org/10.1007/Bf03161886
25. Jeschke G, Pannier M, Spiess HW (2002) Double electron-electron resonance. In: Berliner LJ, Eaton GR, Eaton SS (eds) Distance measurements in biological systems by EPR. Biological magnetic resonance, vol 19. Springer US, New York, pp 493–512. https://doi.org/10.1007/0-306-47109-4_11
26. Bode BE, Margraf D, Plackmeyer J, Durner G, Prisner TF, Schiemann O (2007) Counting the monomers in nanometer-sized oligomers by pulsed electron—electron double resonance. J Am Chem Soc 129(21):6736–6745
27. Jeschke G, Sajid M, Schulte M, Godt A (2009) Three-spin correlations in double electron-electron resonance. Phys Chem Chem Phys 11(31):6580–6591
28. Godt A, Franzen C, Veit S, Enkelmann V, Pannier M, Jeschke G (2000) EPR probes with well-defined, long distances between two or three unpaired electrons. J Org Chem 65 (22):7575–7582
29. Godt A, Schulte M, Zimmermann H, Jeschke G (2006) How flexible are poly (para-phenyleneethynylene)s? Angew Chem Int Edit 45(45):7560–7564. https://doi.org/10.1002/anie.200602807
30. Jeschke G, Sajid M, Schulte M, Ramezanian N, Volkov A, Zimmermann H, Godt A (2010) Flexibility of Shape-Persistent Molecular Building Blocks Composed of p-Phenylene and Ethynylene Units. J Am Chem Soc 132(29):10107–10117
31. Pornsuwan S, Bird G, Schafmeister CE, Saxena S (2006) Flexibility and lengths of bis-peptide nanostructures by electron spin resonance. J Am Chem Soc 128(12):3876–3877
32. Bird GH, Pornsuwan S, Saxena S, Schafmeister CE (2008) Distance distributions of end-labeled curved bispeptide oligomers by electron spin resonance. ACS Nano 2(9):1857–1864. https://doi.org/10.1021/nn800327g
33. Lovett JE, Hoffmann M, Cnossen A, Shutter ATJ, Hogben HJ, Warren JE, Pascu SI, Kay CWM, Timmel CR, Anderson HL (2009) Probing flexibility in porphyrin-based molecular wires using double electron electron resonance. J Am Chem Soc 131(38):13852–13859. https://doi.org/10.1021/ja905796z
34. Jeschke G, Godt A (2003) Co-conformational distribution of nanosized [2] catenanes determined by pulse EPR measurements. ChemPhysChem 4(12):1328–1334
35. Pornsuwan O, Schafmeister CE, Saxena S (2008) Analysis of the dynamical flexibility of bis-peptide nanostructures. J Phys Chem C 112(5):1377–1384. https://doi.org/10.1021/jp077523e
36. Jeschke G, Chechik V, Ionita P, Godt A, Zimmermann H, Banham J, Timmel CR, Hilger D, Jung H (2006) DeerAnalysis2006—a comprehensive software package for analyzing pulsed ELDOR data. Appl Magn Reson 30(3–4):473–498. https://doi.org/10.1007/Bf03166213
37. Wilhelm J, Frey E (1996) Radial distribution function of semiflexible polymers. Phys Rev Lett 77(12):2581–2584
38. Pievo R, Casati C, Franchi P, Mezzina E, Bennati M, Lucarini M (2012) End-to-end distance determination in a Cucurbit[6]uril-Based Rotaxane by PELDOR spectroscopy. ChemPhysChem 13(11):2659–2661
39. Jeschke G, Koch A, Jonas U, Godt A (2002) Direct conversion of EPR dipolar time evolution data to distance distributions. J Magn Reson 155(1):72–82. https://doi.org/10.1006/jmre.2001.2498
40. Polyhach Y, Bordignon E, Tschaggelar R, Gandra S, Godt A, Jeschke G (2012) High sensitivity and versatility of the DEER experiment on nitroxide radical pairs at Q-band frequencies. Phys Chem Chem Phys 14(30):10762–10773
41. Margraf D, Bode BE, Marko A, Schiemann O, Prisner TF (2007) Conformational flexibility of nitroxide biradicals determined by X-band PELDOR experiments. Mol Phys 105(15–16):2153–2160. https://doi.org/10.1080/00268970701724982

42. Savitsky A, Dubinskii AA, Flores M, Lubitz W, Mobius K (2007) Orientation-resolving pulsed electron dipolar high-field EPR spectroscopy on disordered solids: I. Structure of spin-correlated radical pairs in bacterial photosynthetic reaction centers. J Phys Chem B 111 (22):6245–6262. https://doi.org/10.1021/jp070016c
43. Marko A, Margraf D, Yu H, Mu Y, Stock G, Prisner T (2009) Molecular orientation studies by pulsed electron-electron double resonance experiments. J Chem Phys 130(6):064102. https://doi.org/10.1063/1.3073040
44. Marko A, Margraf D, Cekan P, Sigurdsson ST, Schiemann O, Prisner TF (2010) Analytical method to determine the orientation of rigid spin labels in DNA. Phys Rev E 81(2):021911. https://doi.org/10.1103/PhysRevE.81.021911
45. Schiemann O, Cekan P, Margraf D, Prisner TF, Sigurdsson ST (2009) Relative orientation of rigid nitroxides by PELDOR: beyond distance measurements in nucleic acids. Angew Chem Int Edit 48(18):3292–3295
46. Margraf D, Cekan P, Prisner TF, Sigurdsson ST, Schiemann O (2009) Ferro- and antiferromagnetic exchange coupling constants in PELDOR spectra. Phys Chem Chem Phys 11(31):6708–6714
47. Reginsson GW, Hunter RI, Cruickshank PAS, Bolton DR, Sigurdsson ST, Smith GM, Schiemann O (2012) W-band PELDOR with 1 kW microwave power: Molecular geometry, flexibility and exchange coupling. J Magn Reson 216:175–182
48. Bode BE, Plackmeyer J, Prisner TF, Schiemann O (2008) PELDOR measurements on a nitroxide-labeled Cu(II) porphyrin: orientation selection, spin-density distribution, and conformational flexibility. J Phys Chem A 112(23):5064–5073. https://doi.org/10.1021/jp710504k
49. Bode BE, Plackmeyer J, Bolte M, Prisner TF, Schiemann O (2009) PELDOR on an exchange coupled nitroxide copper(II) spin pair. J Organomet Chem 694(7–8):1172–1179. https://doi.org/10.1016/j.jorganchem.2008.11.029

Chapter 4
Other Biradicals and Spin Labeling

Abstract This chapter describes PELDOR or DEER spectroscopic measurements on stable biradicals other than nitroxides and strategies for spin labeling of biomolecules. The most common non-nitroxide biradicals are based on *tris*-(tetrathiaaryl) methyl radicals, known as trityl or TAM radicals. Their narrow spectral widths can present challenges for measurements, but their slow spin relaxation and intense signals make it possible to measure quite long distances. Metal complexes are also used to construct biradicals, particularly ones having moieties with very different spectral characteristics. Many of these same radicals along with nitroxides are used for site-directed spin labelling of biomolecules and complexes, including proteins and nucleic acids.

4.1 Trityl Biradicals

A new class of free radicals, *tris*-(tetrathiaaryl)methyl radicals, commonly called trityl radicals or TAMs, are increasingly used in PELDOR studies. Trityl radicals are structurally and chemically unrelated to nitroxyl radicals. But trityls have some important complimentary properties and are rapidly finding application in chemistry and biology.

The original synthesis of this class of stable trityl radicals was patented in the 1990s, although less-stable triphenyl-methyl radicals have a long history in organic chemistry. The detailed synthesis of stable trityl radicals was provided in 2002 [1], for use in EPR imaging. The structure of a popular trityl radical **4-1**, often called Finland trityl, is shown in Table 4.1.

4.1.1 Trityl Features

Trityl radicals are very suitable for many EPR applications, with chemical derivatives that are soluble in aqueous or non-aqueous solutions. The trityl EPR

© Springer Nature Switzerland AG 2019
Y. D. Tsvetkov et al., *Pulsed Electron–Electron Double Resonance*,
https://doi.org/10.1007/978-3-030-05372-7_4

spectrum generally has one intense, narrow line flanked by a few very weak lines from hyperfine couplings of natural abundance ^{13}C and a transverse relaxation time T_m in the microsecond range even in room-temperature liquids [2, 3]. Trityl radicals are rather stable against reaction with themselves and have in vivo survival times in the range of hours [4, 5].

The trityls have major advantages over nitroxyls for single-frequency pulsed EPR methods, e.g., 2+1, DQC, SIFTER. The narrow trityl EPR linewidth of $\sim$5.6 MHz gives roughly 15-fold greater sensitivity than a nitroxyl with $\sim$70 MHz linewidth at the same radical concentration. An increase in T_m increases the useable duration of the time traces and therefore extends the longest distance limits.

Despite these advantages, trityls pose a problem for PELDOR: with such a small EPR linewidth, it seems difficult to position the pump and observe frequencies to meet the conventional assumptions of PELDOR, see Sect. 1.2. One frequent choice is to observe at the weak shoulders that extend about 15 MHz on either side of the main peak and pump at the main peak. The shoulders are at half the anisotropic ^{13}C hyperfine coupling for the 1-phenyl and 2,6-phenyl carbons [6] and at the position at X-band of spin-flip satellite lines from ^{1}H in the solvent or the trityl [7]. The $\Delta\omega_{AB}/2\pi$ is limited to $\sim$15 MHz, which restricts the ω_1 and pulse powers that can be used. In addition, because the shoulders arise from anisotropic ^{13}C hyperfine couplings, the possibility for orientation selection by the observe pulses must be considered during analysis.

The sharp trityl spectrum that is so valuable for EPR imaging and CW EPR studies means that there is a significant probability that two radicals will have resonant frequencies that differ by less than the dipolar interaction. In that case, the dipolar splitting actually increases by as much as 50% over Eq. 1.2 because pseudosecular terms in the dipolar interaction must be included [8]. This is a problem for single-frequency dipolar spectroscopy methods, e.g., DQC and '2+1', when both trityl spins belong to the sharp central EPR line. However, it ceases to be a problem for PELDOR measurements of distances exceeding 2 nm using the ^{13}C shoulders because with $\Delta\omega_{AB}/2\pi \approx 15$ MHz, only spins with resonance frequency differences larger than the dipolar interaction will contribute to the signal.

Another feature of trityl radicals has raised concerns among a few researchers regarding the analysis of PELDOR data. In nitroxyls, the majority of the spin density is localized in a small volume around the N-O bond [9]. But in trityl radicals, about 30% of the unpaired spin is delocalized across several C atoms at the core of the trityl [6]. The dipole moment of the electron is not focused in a point to the extent it is in nitroxyls, and there is a concern that the point dipole approximation for the dipolar interaction may not hold [9]. However, the trityl core is highly symmetric. To the extent that the trityl's electron spin distribution approaches spherical symmetry, it will behave exactly as a point dipole with any spins outside the trityl core. Recent work with trityl-labeled macromolecules [10–12] and biradicals have not found this concern to be an issue [8, 13–15].

The unpaired spin distribution of trityl radicals seems quite unaffected by solvent or by chemical modification of the trityl [16]. This insensitivity to surroundings makes trityls ideal as labels to report on spin-spin interactions or as tags. However, it does make it difficult to use trityl radicals as spin probes to report on the chemical or physical properties of their surroundings.

4.1.2 Trityl PELDOR

Several studies at X-band compare methods to measure distance: CW EPR, DQC and PELDOR [8, 13–15]. We will focus on the PELDOR results from five types of trityl-containing biradicals, Table 4.1. The arrows in the ESE-detected field-swept EPR spectra of **4-3**, **4-7** and **4-5**, Fig. 4.1, mark the positions of the PELDOR pump and observe pulses.

A pronounced Pake dipolar doublet is seen for **4-3**, Fig. 4.1a, and is quite rare in the EPR literature. This Pake doublet corresponds to a dipolar coupling of 9.5 MHz. The pump pulse was placed at the center of the EPR spectrum of **4-3** for 4pPELDOR measurements and the observe pulse was 5.4 G away in the region of the ^{13}C shoulder for $\Delta\omega_{AB}/2\pi = 15\,\text{MHz}$, Fig. 4.1a. The duration of the pump pulse was 60 ns and the $\pi/2$ and π observe pulses were 32 and 64 ns, respectively, to minimize spectral overlap of the pulses. The excitation width at half-height for the pump and the π observe pulses is about 8 MHz, similar to the 9.5 MHz dipolar coupling of **4-3** determined from CW EPR spectra [8].

The PELDOR time trace $V(T)$ of **4-3** has a clear modulation but a rather poor signal to noise ratio and low modulation amplitude. Its Fourier spectrum has prominent peaks slightly less than 15 MHz rather than the 9.5 MHz expected from the CW EPR spectra. The $V(T)$ was analyzed to give a distance $r = 1.55\,\text{nm}$ between trityl spins [8]. The pseudo-secular part of the dipole–dipole coupling as well as spin delocalization had to be included to reach good agreement with the experimental $V(T)$. The resulting $F(r)$ indicates a distinctly shorter separation between trityls than do modelling or the CW EPR spectra. Interestingly, the ~15 MHz reported dipolar coupling from PELDOR is significantly larger than the 9.5 MHz from CW EPR, but is uncomfortably close to the 14.8 MHz nuclear Zeeman frequency of ^{1}H ESEEM.

Deuterated solvent was used in this measurement to eliminate ^{1}H ESEEM arising from solvent. Yet, each trityl groups contains 36 ^{1}H in its methyl groups and another 10 in the ethyl esters. The spin-flip satellite line intensity and ESEEM intensity was found to be roughly the same for the methyl ^{1}H as for nearby solvent ^{1}H in the very similar free-acid of Finland trityl [7]. There is no mention that averaging was used to remove ^{1}H nuclear modulation [8], so the $V(T)$ analysis and the resultant $F(r)$ of **4-3** may be artifacts from ^{1}H ESEEM from the biradical itself and not from any dipolar interaction.

The much longer trityl biradical **4-7** was studied with the same measurement conditions used for **4-3**, except that the observe pulses were directly on the ^{13}C satellites, 4.6 G from the pump pulse with $\Delta\omega_{AB}/2\pi = 13\,\text{MHz}$, Fig. 4.1b [8]. This $V(T)$ has strong, clear modulation which is closely fit by the simulated time trace,

Table 4.1 Structure and distance data for Finland trityl, its diethyl ester (T) and some simple biradicals, Et = ethyl, Hept = heptyl

	Biradical or radical	r, nm	References
4-1	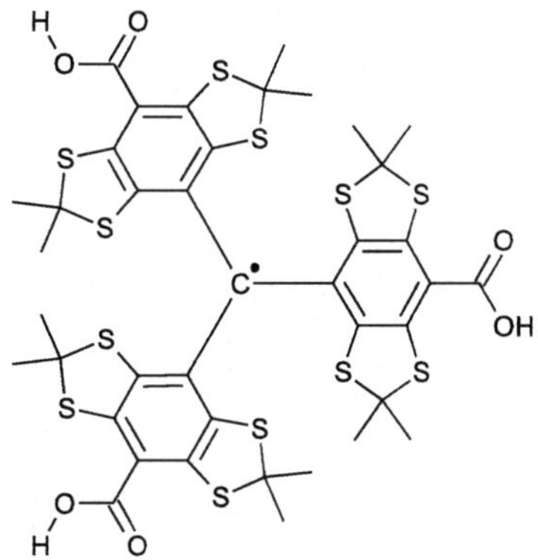	N.A.	

(continued)

Table 4.1 (continued)

	Biradical or radical	r, nm	References
4-2	T= 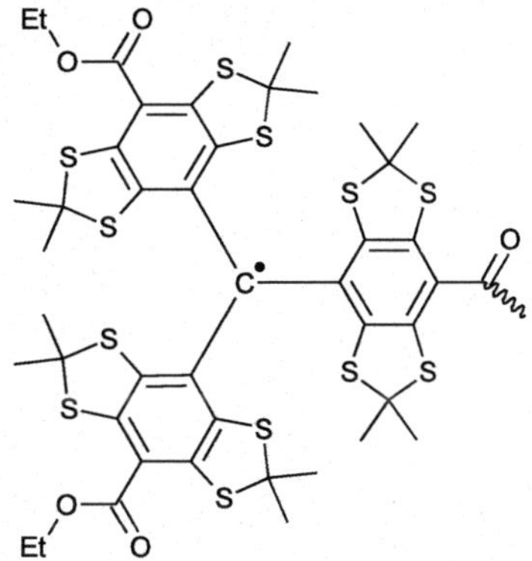	N.A.	[1]
4-3	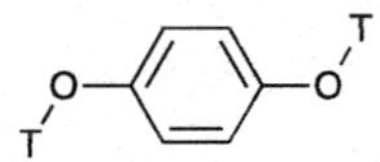	1.55 ± 0.06	[8]
4-4	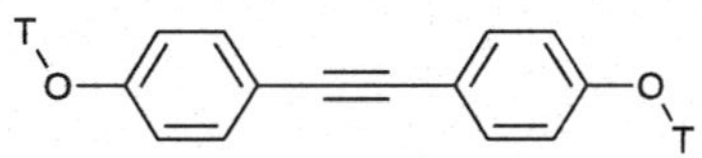	2.17 ± 0.08[a]	[8]

(continued)

Table 4.1 (continued)

	Biradical or radical	r, nm	References
4-5	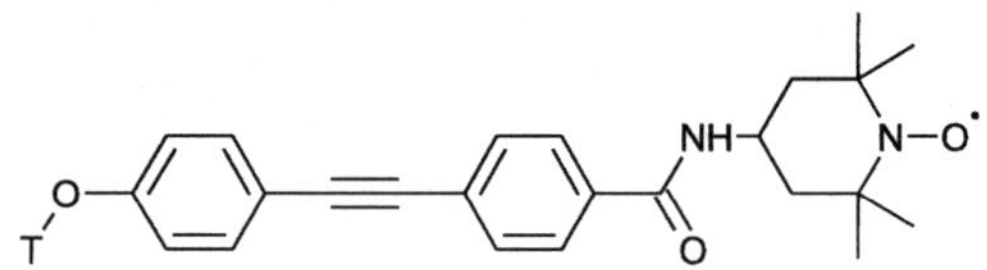	2.43 ± 0.16	[8]
4-6	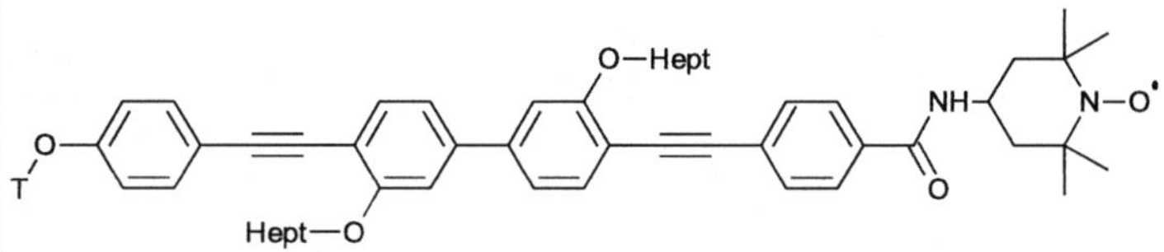	3.48 ± 0.11	[13]
4-7	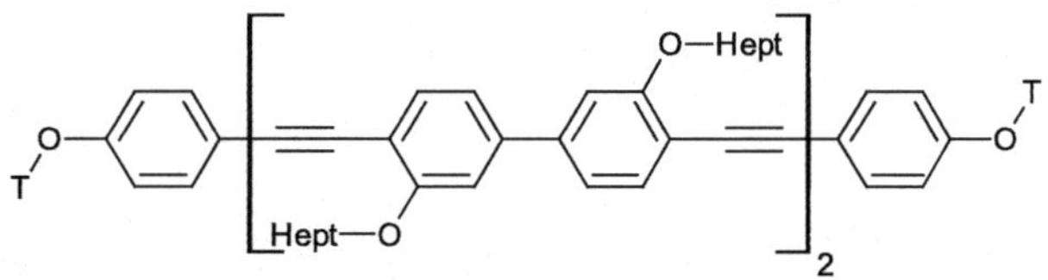	4.89	[13]

[a]Distance from DQC measurements

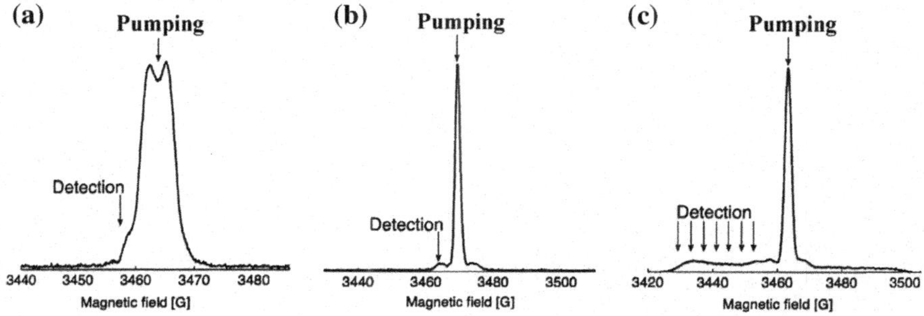

Fig. 4.1 Two pulse ESE-detected field sweeps of **a 4-3**, **b 4-7** and **c 4-5**; arrows show the positions of the PELDOR pump and observe pulses; the original paper appears to exchange **4-7** and **4-5**. Reprinted from Kunjir et al. [8] with permission of Royal Society of Chemistry, copyright 2013

Fig. 4.2a. Tikhonov regularization gave a $F(r)$ with a mean 4.90 nm distance [8], which agrees well with a rigid-segment model of **4-7**, Fig. 4.2b [13].

The PELDOR measurements on **4-7**, and possibly **4-3**, show that it is possible to use PELDOR to measure the dipolar coupling between trityl radicals. However, the narrow spectral width of the trityl radical at X-band requires PELDOR pulses with a narrow excitation bandwidth, which limits the p_A and p_B pulse parameters, signal to noise ratio, and the distance range accessible to PELDOR measurement within the conventional assumptions for PELDOR, Eqs. 1.6–1.9. Trityl radical derivatives with somewhat broader EPR spectra or with two lines may be better suited for PELDOR measurements.

A hybrid nitroxide-trityl biradical has some of the advantages of the sharp trityl spectrum with greater choice of pump offsets $\Delta\omega_{AB}$. This is seen in PELDOR measurements of the interspin distance of **4-5**. The pump pulse was placed on trityl at the maximum of the spectrum, Fig. 4.1c, to maximize p_B and depth of the modulation in $V(T)$. The observe pulses were placed on the nitroxyl lines 30–90 MHz higher in frequency. The $V(T)$ of **4-5** have great modulation depth, about 80%. The modulation frequency depends on the position of the observe pulses in the spectrum, indicating orientation selection, and that the nitroxyl spin label's orientation is correlated with the interspin vector.

An orientation-averaged $V(T)$ was obtained by adding together individual time traces, Fig. 4.3b. Analyzing the averaged $V(T)$ with Tikhonov regularization yields a major $F(r)$ peak with a mean value of 2.43 nm, Fig. 4.3b inset. A smaller peak, at 1.9–2.0 nm, is due to incomplete orientation averaging.

The most accurate $F(r)$ for **4-5** was obtained by simulating the $V(T)$ using the rigid-segment model, including the secular part of the dipole–dipole coupling and spin delocalization for the trityl radical, Fig. 4.3b. PELDOR and DQC measurements agree completely for biradicals **4-3–4-6** [8].

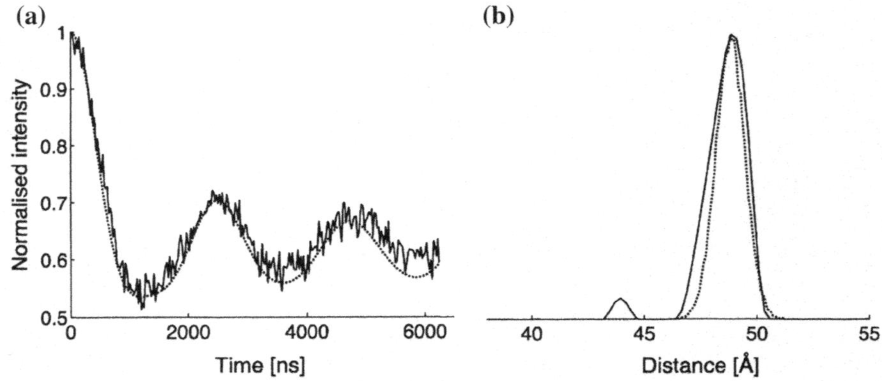

Fig. 4.2 PELDOR measurements of compound **4-7**: **a** background corrected experimental PELDOR time trace of **4-7** (solid line) and the simulated time trace obtained from a rigid segment model of **4-7** (broken line), **b** distance distributions from the rigid-segment model of **4-7** (broken line) and Tikhonov analysis on the PELDOR time trace (solid line). Reprinted from Kunjir et al. [8] with permission of Royal Society of Chemistry, copyright 2013

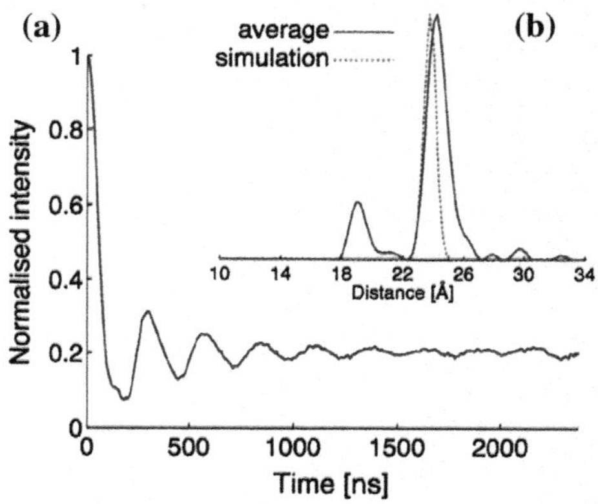

Fig. 4.3 PELDOR data from **4-5**: **a** Orientation-averaged PELDOR time trace, **b** $F(r)$ for **4-5** from the orientation-averaged PELDOR time trace and from the rigid segment model. Reprinted from Kunjir et al. [8] with permission of Royal Society of Chemistry, copyright 2013

4.1.3 Trityls at High Frequency

Pulsed EPR dipolar spectroscopy measurements at Q- and G-band were made on the long trityl biradical **4-7**, Table 4.1 [14]. The most interest results were at G-band where the EPR spectrum has a spectral width $\Delta\omega/2\pi \sim 130$ MHz and pronounced asymmetry due to partially-resolved axial g-tensor anisotropy. The width of the trityl EPR spectrum scales with mw frequency, and is $\sim$55 MHz at Q-band, because the primary contribution to the width is g-tensor anisotropy. Normalized G-band PELDOR time traces were obtained with pump and observe pulse positions

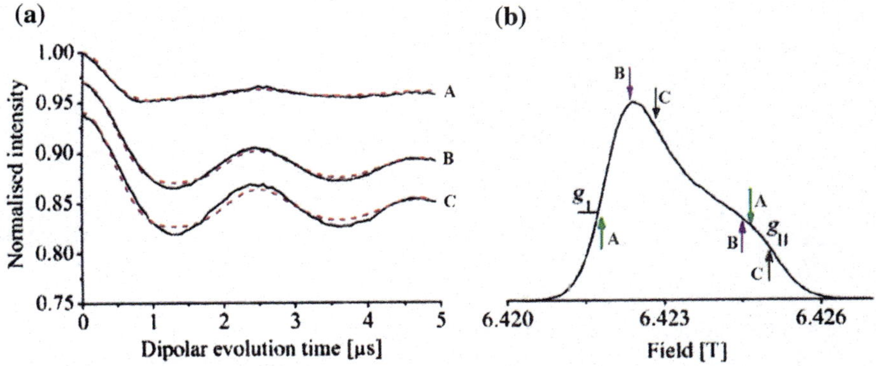

Fig. 4.4 **a** Normalized PELDOR time traces of **4-7** at G-band, solid lines are experimental data, dashed lines are fits using Tikhonov regularization, traces *A*, *B* and *C* were measured at the positions indicated in the spectrum on the right; the $\Delta\omega_{AB}/2\pi$ is 80 MHz for *A* and 60 MHz for *B* and *C*; the traces obtained at different positions are vertically shifted; **b** Field swept ESE-detected EPR spectrum at G-band at 25 K upward arrows indicating observe positions and downward arrows pump positions. Reprinted from Akhmetzyanov et al. [69] with permission of Royal Society of Chemistry, copyright 2013

A, *B* and *C*, Fig. 4.4b. At position *A*, the observe pulses excite primarily at the perpendicular components of the *g*-tensor, $g_\perp$, of each trityl. Conversely at *B* and *C*, the observe pulses excite predominantly at $g_\parallel$. However, the relatively small *g*-tensor anisotropy makes orientation selection effects less pronounced than for nitroxyl rigidly attached to biomolecules. For this reason, the PELDOR time traces do not provide unambiguous orientation information for this molecule.

Nevertheless, the weak orientation selection effects can be understood in terms of the molecular structure and *g*-tensor orientation. The $g_\parallel$ of **4-7** is perpendicular to the mean linker axis of the molecule. Therefore, when the $g_\parallel$ region of the spectrum is observed in traces B and C, the perpendicular part of the dipolar coupling tensor is seen, while A sees parallel and perpendicular orientations of the dipolar coupling tensor. Tikhonov analysis of the time traces from *A* give a mean distance at about 4.92 nm and about 4.88 nm from *B* and *C*.

A multi-frequency approach has great potential for biological applications of trityl spin tags for nucleic acids or proteins in cells under physiological conditions.

4.2 Paramagnetic Ion-Nitroxyl Pairs

The nitroxyl biradicals discussed in Chap. 3 provide unique information about molecular structure, flexibility and electron exchange. But they also have had an important role as simple, well-defined models for biomacromolecules with nitroxyl spin labels attached. Many important proteins, enzymes, and nucleic acids contain, or can form, complexes with paramagnetic metals, so that attachment of a nitroxyl

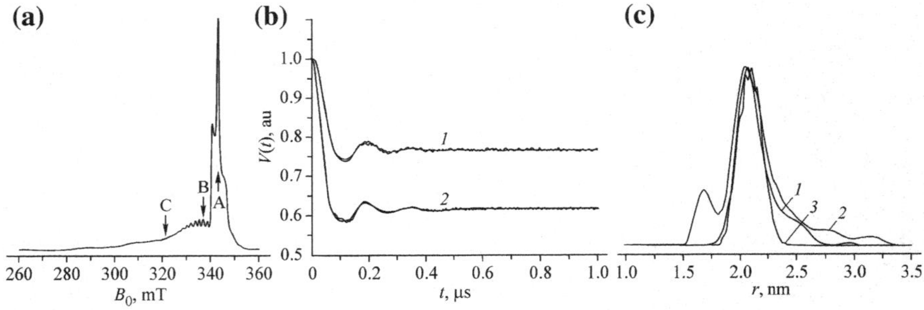

Fig. 4.5 **a** EPR spectrum of **4-8** measured by the two-pulse ESE signal at X-band, arrows indicate PELDOR observe (*A*) and pump (*B* or *C*) locations; **b** 1- PELDOR time traces for $\Delta\omega_{AB}/(2\pi)$ = 603 MHz and 2- for $\Delta\omega_{AB}/(2\pi)$ = 226 MHz.; **c** $F(r)$ from Tikhonov regularization 1- for $\Delta\omega_{AB}/(2\pi)$ = 226 MHz, 2- for $\Delta\omega_{AB}/(2\pi)$ = 603 MHz, 3- calculated for a Gaussian distribution of copper ion–nitroxyl distances in the porphyrin ring plane. Reprinted from Bode et al [17] with permission of American Chemical Society, copyright 2008

spin-label produces a spin pair composed of a paramagnetic ion and a nitroxyl radical.

Such pairs are an attractive subject for PELDOR studies because metal ions play a key role in the biological activity of many proteins and enzymes. A few molecular constructs, Table 4.2, that model pairs composed of a paramagnetic metal ion and a nitroxyl radical have been studied by PELDOR. These molecular systems are also of interest from the standpoint of developing artificial supramolecular ensembles with specified properties. The metal ion alters some of the paramagnetic properties of such pairs. Their EPR spectra are broader than those of nitroxyl biradicals because of the *g*-factor anisotropy of the metal, Fig. 4.5a, so that orientation selection is a major consideration for PELDOR studies involving most paramagnetic metal ions, Sects. 1.4 and 3.3. The range of distances accessible by PELDOR is typically smaller than for nitroxyl radicals because the ions often shorten the phase and spin-lattice relaxation times in ion–nitroxyl pairs. But, PELDOR measurements are possible with many ions at temperatures where the relaxation times are reasonable, e.g., with Cu^{2+} and Gd^{3+}.

4.2.1 Distance

Paramagnetic copper (II) ions are present in many metallo-proteins, biological systems, and supramolecular ensembles. An $S = 1/2\,Cu^{2+}$ porphyrin linked to a nitroxyl radical, **4-8**, was studied in detail by PELDOR [17]. For each molecule, the PELDOR time traces for $\Delta\omega_{AB}/(2\pi) = 226$ or 603 MHz, spectral positions *B* and *C*, respectively, are slightly different, Fig. 4.5. These traces come from the spin dipole interaction, but the delocalization of spin density from the copper onto the nitrogen atoms must be considered, along with orientation selection and the conformational flexibility of the molecules [17].

Tikhonov regularization for $V(T)$ analysis assumes no orientation selection, Sect. 1.4. Artifacts arise when Tikhonov regularization is applied to metal ion data with even moderate orientation selection, as seen with **4-8**, Fig. 4.5c [17]. The main $F(r)$ peak for both values of $\Delta\omega_{AB}$ coincide at $r = 2.06$ nm and agree with X-ray diffraction data. However, $F(r)$ has shoulders and several false maxima at smaller and larger r values, which can lead to false conclusions about ion–radical conformations. Analysis of $V(T)$ taking orientation selection into account has just a single Gaussian peak, Fig. 4.5c [17]. When the mutual orientation of spin fragments is fixed rigidly, as it is in proteins, i.e., giving strong orientation selection, the solution of the inverse problem by standard Tikhonov regularization is incorrect and gives significant artifacts.

A six-coordinate Cu^{2+} complex **4-10** containing two nitroxyl radicals was studied at 15 K at X-band [19]. The EPR spectrum contains a typical nitroxyl triplet with $g \approx 2.003$ and an anisotropic line, ~55 mT wide, from the $S = 1/2$ Cu^{2+} ion with $g_z = 2.258$. PELDOR $V(T)$ were measured using 32 ns pulses with the pump pulse fixed on the nitroxyl maximum. When the observe pulses excited Cu^{2+} with $\Delta\omega_{AB}/(2\pi)$ of 200 MHz or 1465 MHz, the $V(T)$ yielded distances between >N–O$^{\cdot}$ and Cu^{2+} of $r = 2.43$ nm and 2.46 nm, respectively. When the nitroxyl radical was observed, $V(T)$ gave a distance of 5.2 nm between the nitroxyl groups of the complex, agreeing well with calculations [19].

The Cu^{2+} was replaced by a high-spin Gd^{3+} ion, with $S = 7/2$, to make **4-11** [20]. These ions have slow relaxation in frozen deuterium-containing solutions, comparable to those for nitroxyls, and weak orientation selection. At X-band, the nitroxyl EPR spectrum overlaps the broad EPR peak of the $-1/2 \rightarrow +1/2$ transition of Gd^{3+}, but at Q-band this Gd^{3+} peak lies outside the nitroxyl triplet. The pump pulse was applied at the maximum of the nitroxyl spectrum, while the observe pulse was 80 MHz away at X-band or 300 MHz at Q-band. The PELDOR sensitivity at Q-band was 34-fold greater than at X-band [20]. At both bands, $V(T)$ modulation of **4-11** gave $r = 2.54$ nm, coinciding with **4-10**.

However, the modulation depth at Q-band is as large as at X-band. The Tikhonov analysis of $V(T)$ reveals satellites near 2.8 nm at X-band and 1.8 nm at Q-band, in addition to the main line at 2.54 nm. These satellite lines correspond to deuterium ESEEM modulation. Several hypotheses were considered, but their origin remains uncertain [20]. Despite difficulties in interpretation, Gd^{3+} ions may be a valuable alternative to nitroxyl labels for PELDOR studies of biologically important molecules, especially at Q-band [20].

Table 4.2 Complexes studied by PELDOR containing copper-octaethyl porphyrin or terpyridine complexes linked to a nitroxide, Hex = hexyl

Complexes		References
4-8	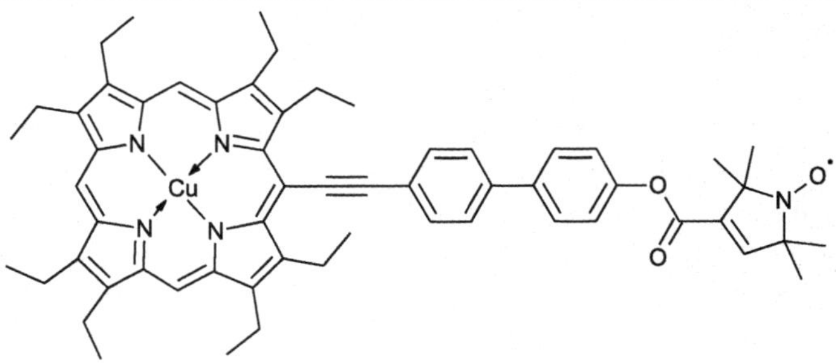	[17, 18]
4-9	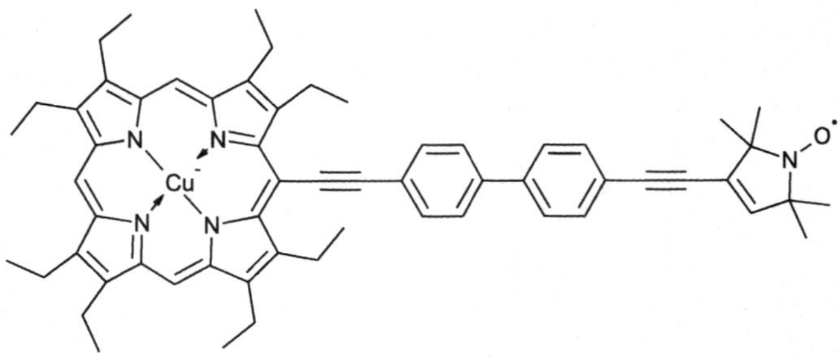	[18]
4-10	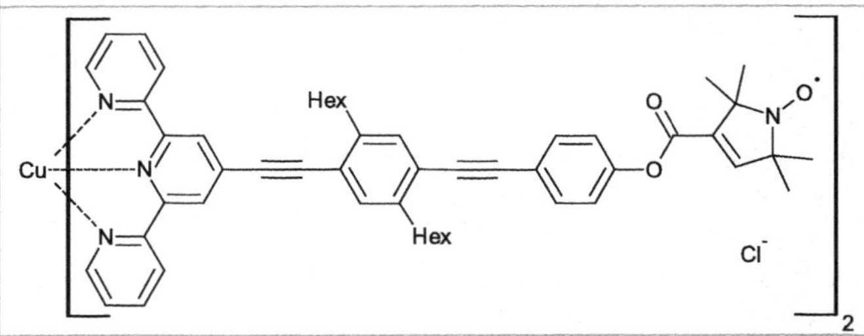	[19]
4-11	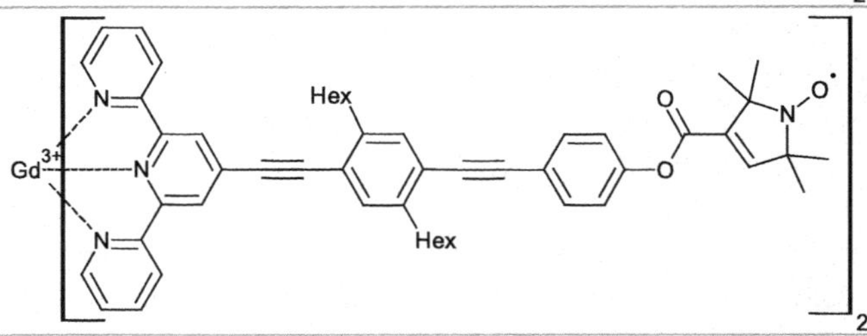	[20–22]

4.3 Site-Directed Spin Labeling

Site-directed spin labeling SDSL methods opened new vistas for EPR spectroscopy in biochemistry and biophysics, e.g., determination of the secondary and tertiary structure of proteins in solution and in membranes, investigation of the orientation and motion of particular protein domains under physiological conditions, determination of conformational transitions in functioning protein-membrane systems. Initially, these investigations were carried out by CW EPR spectroscopy and are systematically reviewed in the book series *Biological Magnetic Resonance* edited by L. Berliner.

Briefly, SDSL uses molecular biology techniques to produce proteins with specific amino acid residues, typically cysteine, placed at desired sites in the protein sequence followed by attachment of spin labels to those sites. SDSL techniques readily produce doubly spin labeled proteins for PELDOR studies of a wide range of biomolecules. The SDSL concept has been extended to site specific labeling of nucleic acids at single or multiple sites. Some of the popular labels, Tables 4.3 and 4.4, and the methods for their attachment to biomolecules are given below. Vols. 14 and 21 in the *Biological Magnetic Resonance* series [23, 24] have excellent reviews of the pioneering studies applying PELDOR to biomolecules.

4.3.1 Labels for Proteins

Many different spin labels have been developed for proteins. The structures of a few commonly-used labels, are shown in Table 4.3. Some labels are known by multiple abbreviations and some abbreviations are applied to multiple labels. Structures **4-12** to **4-15** show the spin label attached to the sulfur of cysteine, **4-16–4-18**, are spin label reagents before attachment to a protein and **4-19** is a common, 6-membered nitroxyl ring used in many labels and biradicals.

Virtually any bioconjugate chemistry can be used to attach a spin label to a protein, but the field has converged to the labeling of cysteine because cysteine is a low-abundance amino acid which can be labeled under very mild conditions in high yield with few side reactions. The development of efficient SDSL methods for biologically valuable compounds was the key step which inevitably drove the application of PELDOR to biomolecules.

The most popular method of labeling proteins is to use standard genetic engineering methods to replace all cysteine amino acids in the protein; place cysteines where the spin labels are desired; express and purify the protein; and attach a nitroxyl to each cysteine using a radical-substituted thiosulfonate **4-12** via reaction A [25–28]. The labeled protein is then recovered for study. In many cases, the labeled protein has the same structure, behavior and properties as the native protein.

Table 4.3 Structure and abbreviations of a few popular spin labels

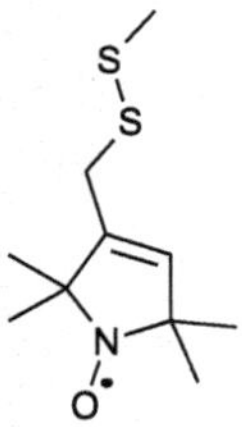	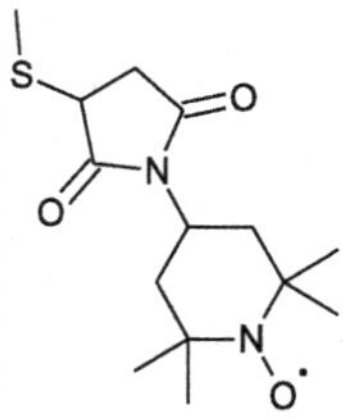	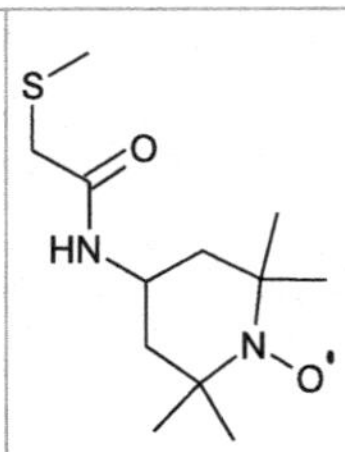	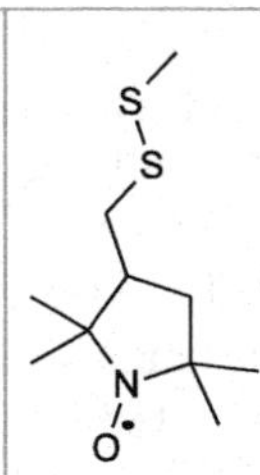
MTS **4-12**	MSL **4-13**	IASL **4-14**	SAT **4-15**
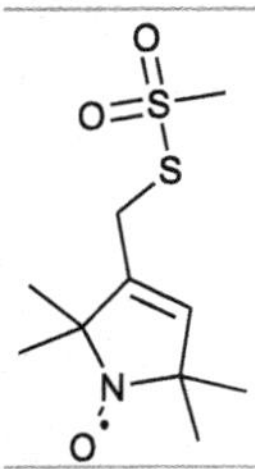	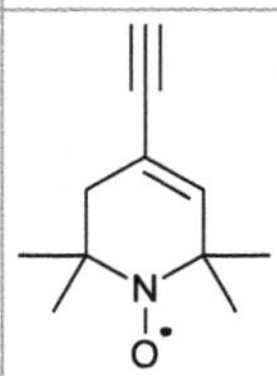	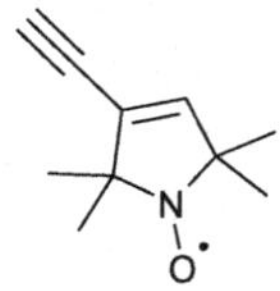	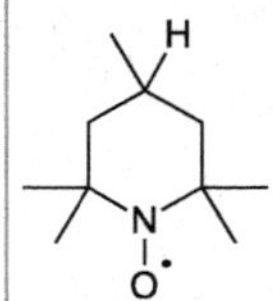
MTSL **4-16**	TEMPA **4-17**	TPA **4-18**	TEMPO **4-19**

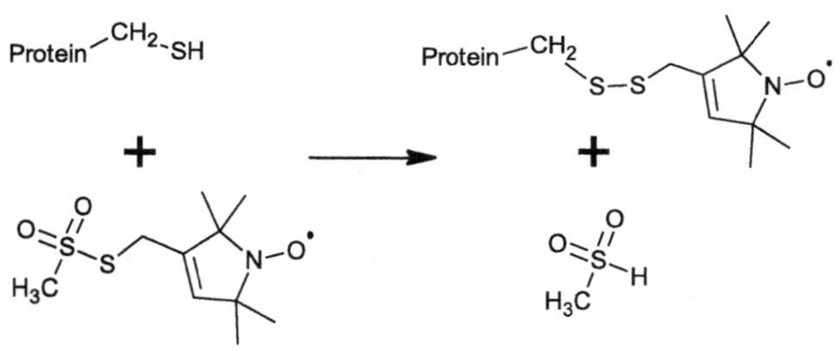

Reaction A

Multiple conformations of the linker connecting the protein and the label cause some ambiguity in the interpretation of PELDOR measurements. Different conformations of the linker makes the position of the cysteine relative to its label somewhat uncertain. The problem is particularly severe for distances of 1.0–2.0 nm between spins, which are not large compared to the size of the linker. Attempts to allow for conformational and structural effects in distance calculations from data on dipolar interactions are made [29]. The best correlations with $R^2 = 0.8$ between EPR-measured distances and calculated distance were obtained using MD calculations to find the lowest-energy conformers [29]. In the case of flexible spin labels, the effect of conformations is usually to broaden the $F(r)$ extracted directly from the PELDOR time trace by Tikhonov regularization methods.

Table 4.4 Structure and abbreviations of a few rigid spin labels

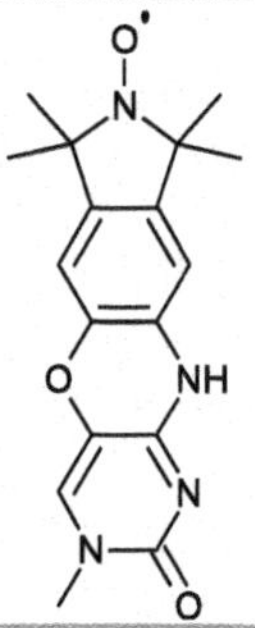	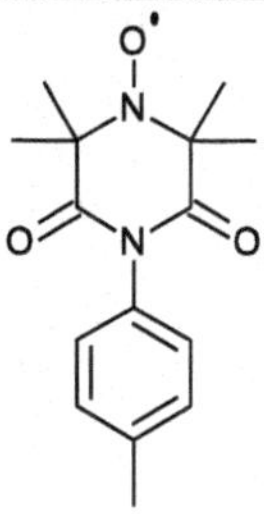	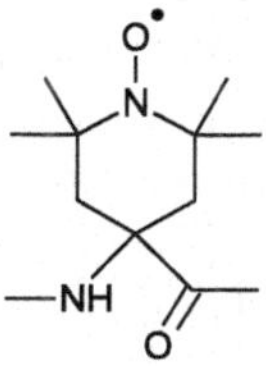	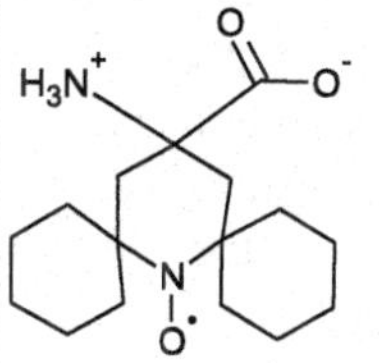	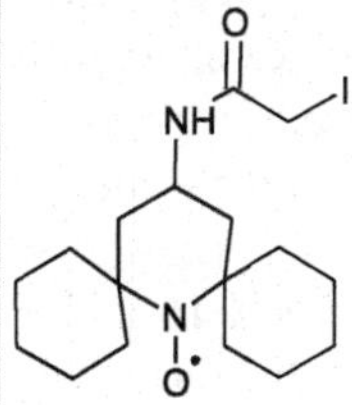
Ç **4-20**	TOPP **4-21**	TOAC **4-22**	**4-23**	**4-24**

A number of attempts have been made to reduce the flexibility and range of conformations that the spin label can adopt in the protein, Table 4.4. The TOAC label **4-22**, and its close relative **4-23**, have some valuable properties as compared to the labels in Table 4.3. TOAC itself is an amino acid that can be incorporated into the backbone of the protein. The paramagnetic nitroxyl probe is physiclly stabilized by two adjacent tetrasubstituted carbon atoms that are in the same six-membered ring as the C_α atom in the protein backbone. This geometry dramatically restricts the flexibility of the nitroxyl spin label and prevents its rotation about side-chain bonds, although the six-membered ring has at least two conformations [30, 31].

Thus, TOAC **4-22** exhibits quite a restricted mobility, much different from all other spin-labeled amino acids that contain a flexible link joining the radical to the peptide backbone. TOAC is a very useful tool in peptide and protein chemistry, which allows incorporation of a probe sensitive to structure and environment. TOAC **4-22** can be prepared on a large scale from commercially-available 2,2,6,6-tetramethyl-4-oxo-piperidine [32, 33]. Unfortunately, TOAC requires rather complex methods to incorporate it into peptides by solid-phase synthesis and to preserve or regenerate its crucial free radical [34, 35].

4.3.2 Labels for Physiological Temperatures

One drawback of PELDOR for biological applications has been the near-impossibility of measurements at physiological temperatures. This difficulty is due to the short phase relaxation time T_M in this temperature range for the commonly-used nitroxyl spin labels. Below $\sim$100 K, the ESE decay for nitroxyl radicals is dominated by nuclear spin diffusion among the solvent nuclei [36]. As the temperature is increased above $\sim$100 K, rotation of methyl groups in methyl-containing nitroxides becomes a highly-effective dephasing mechanism

[37]. The resulting decrease in T_M above $\sim$100 K imposes a practical temperature limit for PELDOR measurements with methyl-containing nitroxyl labels.

Replacing the *gem*-dimethyl groups with spirocyclohexyl groups eliminates this methyl group effect and increases T_M at temperatures $T > 100\,\text{K}$. In order to eliminate methyl groups, TOAC **4-22** was modified to **4-23** [38] and **4-24** [39]. In comparable conditions, the phase relaxation time increased 3-4 times compared with **4-16** and PELDOR measurements became possible at 125-160 K.

In glassy or frozen solvents, the softening of the matrix with increasing temperature increases motion of the labels and averages *g*- and hyperfine anisotropy, which also decreases T_M. In water/glycerol, toluene, alcohol and other solutions, these motions cause the further decrease in T_M above $\sim$180 K. But matrices with higher melting points, e.g., sucrose, trehalose or sorbitol, remain relatively rigid to substantially-higher temperature, and the T_M decreases more gradually. Another way to immobilize a labeled molecule is to use a sorbent. Approaches for PELDOR measurements at ambient temperatures have been recently reviewed [12].

The spirocyclohexyl label **4-24** was attached to sites on T4 lysozyme and allowed interspin distances up to $\sim$4 nm to be measured by PELDOR at temperatures up to 160 K in water/glycerol glasses. In a glassy trehalose matrix, the T_M was long enough with **4-24** to measure an interspin distance of 3.2 nm at 295 K, which was not possible for the same protein labeled with the conventional MTSL **4-16** [39].

Trityl radicals provide even greater prospects for distance measurements at physiological temperatures. Their long intrinsic T_M at room temperature remained even after their rigid attachment to DNA either as an end label or by incorporation in place of a nucleotide. Distances of 5–6 nm were measured at 298 K with the labeled DNA immobilized in trehalose [40].

4.3.3 Labels for Nucleic Acids

Methods to label nucleic acids and oligonucleotides have been slower to develop than those for proteins. However, there are several reviews of nucleic acid spin labeling [23, 24, 41]. Because conformational mobility of the spin labels impacts the accuracy in distance determination by PELDOR, the use of rigid linkers to attach spin labels to nucleic acids has become common. The Sonogashira reaction, which replaces an iodine in an organic compound by an alkynyl group, catalyzed by Pd(II) and Cu(I) salts, is a common approach today [42, 43]. Nucleic acids are labeled by **4-17** or **4-18** at the C5 position of iodouridine via reaction B [28, 44–47] to give adducts where Rib is a ribosyl or deoxyribosyl group.

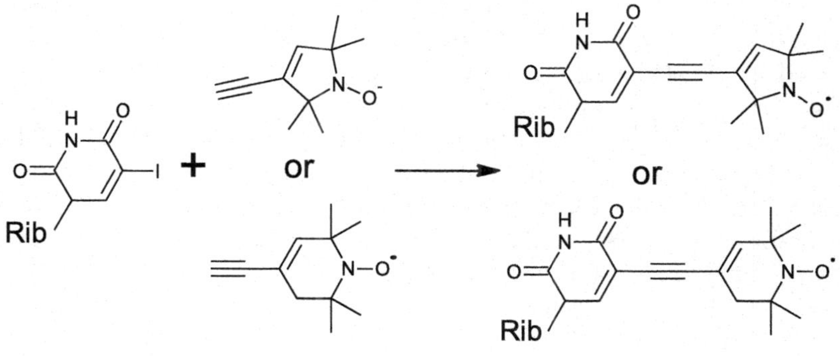

Reaction B

This technique can produce spin-labeled phosphoramidite monomers for oligonucleotide synthesis or it can be used to attach the label into previously-made ribo- and deoxyribo-oligonucleotides containing iodouridine. Adenine and cytosine residues with protected amino groups have also been labeled with **4-18** via the Sonogashira reaction [48].

Azide-alkyne cycloaddition, known as "click chemistry," is also commonly used to introduce spin labels into oligonucleotides [49]. The spin label is attached via reaction C, the Cu(I)-catalyzed reaction of 4-azido-TEMPO with an acetylene group at C7 of 7-deazaadenine or at C5 of uracil [50], or during solid-phase synthesis of oligonucleotides [51]. This chemistry is stereospecific, has a high yield, and works well for both DNA and RNA.

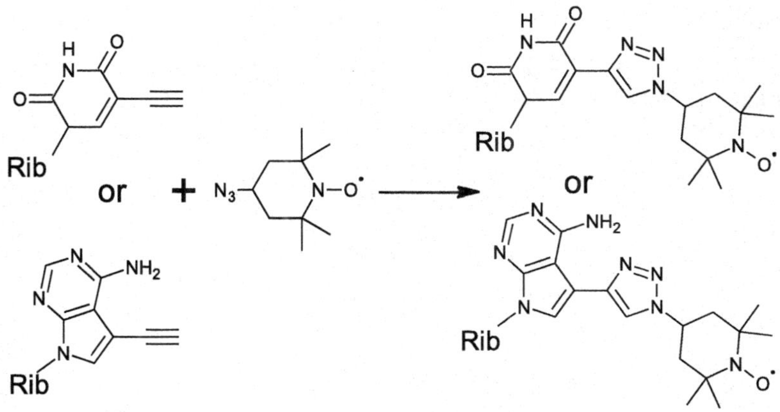

Reaction C

If orientation selection is desired for determination of the specific mutual orientation of the labels along with distance, then rigid spin labels should be used. One is the so-called Ç or "C-spin" label **4-20**, Table 4.4, [52] used for DNA and RNA labeling [53–55]. This label forms a rigid structure by base-pairing to guanine and was used for orientation selection measurements.

Trityl labels have also been used for end labeling of DNA duplexes. The trityl actually seems to form a fairly compact structure with the terminus of the duplex that can make the distance distribution quite narrow [11, 56]. Rigid phosphoramidite trityl labels that can be incorporated into synthetic nucleic acids have also been developed [40]. They are bulky, but do allow measurements of large 5–6 nm distances at physiological temperatures **4.3.2**.

4.4 Other Labels

Nitroxyl spin labels are by far the most common in PELDOR research. However, there is a substantial amount of research done with other paramagnetic centers and naturally-occurring cofactors. For example photosynthetic reaction centers [57], tyrosylamino acid radicals [58, 59], trityl radicals [10, 11], metal centers [60, 61], iron-sulfur clusters [62, 63], or rare earth metal tags [64–66]. A recent review article focuses on spin labels, their synthesis, and spin labeling [67].

A gadolinium chelate Gd^{3+}-DOTA **4-25** has recently been used and is a promising spin label in PELDOR spectroscopy [21, 68]. In harsh environments, **4-25**, Table 4.5, and related compounds are rather stable. They have several advantages especially at Q-, W-, and higher-frequency bands: no orientation selection; higher transition moments owing to their $S = 7/2$ spin state; use of high repetition rates due to the short longitudinal relaxation T_1. These advantages result in greater signal-to-noise-ratios in PELDOR experiments than standard nitroxyl labels [68].

Table 4.5 A Gd^{3+}-DOTA based spin label

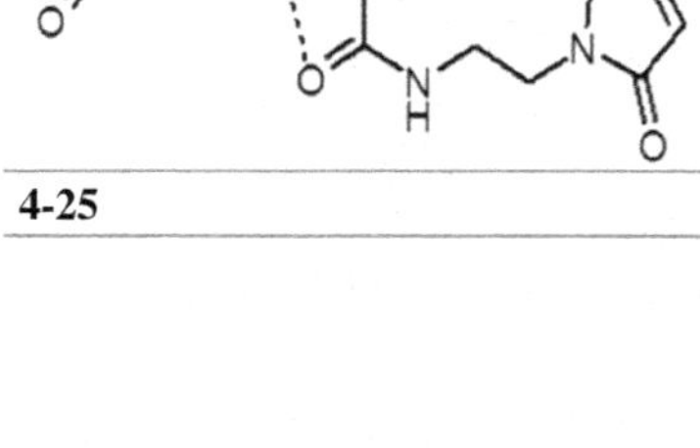

4-25

References

1. Reddy TJ, Iwama T, Halpern HJ, Rawal VH (2002) General synthesis of persistent trityl radicals for EPR imaging of biological systems. J Org Chem 67(14):4635–4639
2. Fielding AJ, Carl PJ, Eaton GR, Eaton SS (2005) Multifrequency EPR of four triarylmethyl radicals. Appl Magn Reson 28(3–4):231–238
3. Owenius R, Eaton GR, Eaton SS (2005) Frequency (250 MHz to 9.2 GHz) and viscosity dependence of electron spin relaxation of triarylmethyl radicals at room temperature. J Magn Reson 172(1):168–175
4. Bobko AA, Dhimitruka I, Zweier JL, Khramtsov VV (2007) Trityl radicals as persistent dual function pH and oxygen probes for in vivo electron paramagnetic resonance spectroscopy and imaging: concept and experiment. J Am Chem Soc 129(23):7240–7241. https://doi.org/10.1021/ja071515u
5. Liu Y, Villamena FA, Sun J, Xu Y, Dhimitruka I, Zweier JL (2008) Synthesis and characterization of ester-derivatized tetrathiatriarylmethyl radicals as intracellular oxygen probes. J Org Chem 73(4):1490–1497
6. Bowman MK, Mailer C, Halpern HJ (2005) The solution conformation of triarylmethyl radicals. J Magn Reson 172(2):254–267
7. Trukhan SN, Yudanov VF, Tormyshev VM, Rogozhnikova OY, Trukhin DV, Bowman MK, Krzyaniak MD, Chen H, Martyanov ON (2013) Hyperfine interactions of narrow-line trityl radical with solvent molecules. J Magn Reson 233:29–36. https://doi.org/10.1016/j.jmr.2013.04.017
8. Kunjir NC, Reginsson GW, Schiemann O, Sigurdsson ST (2013) Measurements of short distances between trityl spin labels with CW EPR, DQC and PELDOR. Phys Chem Chem Phys 15(45):19673–19685. https://doi.org/10.1039/c3cp52789a
9. Karoui H, Moigne FL, Ouari O, Tordo P (2010) Nitroxide radicals: properties, synthesis and applications. In: Stable Radicals. Wiley, pp 173–229. https://doi.org/10.1002/9780470666975.ch5
10. Shevelev GY, Krumkacheva OA, Lomzov AA, Kuzhelev AA, Rogozhnikova OY, Trukhin DV, Troitskaya TI, Tormyshev VM, Fedin MV, Pyshnyi DV, Bagryanskaya EG (2014) Physiological-temperature distance measurement in nucleic acid using triarylmethyl-based spin labels and pulsed dipolar EPR spectroscopy. J Am Chem Soc 136(28):9874–9877. https://doi.org/10.1021/ja505122n
11. Shevelev GY, Krumkacheva OA, Lomzov AA, Kuzhelev AA, Trukhin DV, Rogozhnikova OY, Tormyshev VM, Pyshnyi DV, Fedin MV, Bagryanskaya EG (2015) Triarylmethyl labels: toward improving the accuracy of EPR nanoscale distance measurements in DNAs. J Phys Chem B 119(43):13641–13648
12. Krumkacheva O, Bagryanskaya E (2017) EPR-based distance measurements at ambient temperature. J Magn Reson 280:117–126. https://doi.org/10.1016/j.jmr.2017.02.015
13. Reginsson GW, Kunjir NC, Sigurdsson ST, Schiemann O (2012) Trityl radicals: spin labels for nanometer-distance measurements. Chemistry 18(43):13580–13584. https://doi.org/10.1002/chem.201203014
14. Akhmetzyanov D, Schops P, Marko A, Kunjir NC, Sigurdsson ST, Prisner TF (2015) Pulsed EPR dipolar spectroscopy at Q- and G-band on a trityl biradical. Phys Chem Chem Phys 17(37):24446–24451
15. Yang ZY, Liu YP, Borbat P, Zweier JL, Freed JH, Hubbell WL (2012) Pulsed ESR dipolar spectroscopy for distance measurements in immobilized spin labeled proteins in liquid solution. J Am Chem Soc 134(24):9950–9952
16. Kuzhelev AA, Tormyshev VM, Rogozhnikova OY, Trukhin DV, Troitskaya TI, Strizhakov RK, Krumkacheva OA, Fedin MV, Bagryanskaya EG (2017) Triarylmethyl radicals: EPR study of ^{13}C hyperfine coupling constants. Z Phys Chem 231(4):777–794. https://doi.org/10.1515/zpch-2016-0811

17. Bode BE, Plackmeyer J, Prisner TF, Schiemann O (2008) PELDOR measurements on a nitroxide-labeled Cu(II) porphyrin: orientation selection, spin-density distribution, and conformational flexibility. J Phys Chem A 112(23):5064–5073. https://doi.org/10.1021/jp710504k
18. Bode BE, Plackmeyer J, Bolte M, Prisner TF, Schiemann O (2009) PELDOR on an exchange coupled nitroxide copper(II) spin pair. J Organomet Chem 694(7–8):1172–1179. https://doi.org/10.1016/j.jorganchem.2008.11.029
19. Narr E, Godt A, Jeschke G (2002) Selective measurements of a nitroxide-nitroxide separation of 5 nm and a nitroxide-copper separation of 2.5 nm in a terpyridine-based copper(II) complex by pulse EPR spectroscopy. Angew Chem Int Edit 41(20):3907–3910
20. Lueders P, Jeschke G, Yulikov M (2011) Double electron-electron resonance measured between Gd^{3+} ions and nitroxide radicals. J Phys Chem Lett 2(6):604–609. https://doi.org/10.1021/jz200073h
21. Raitsimring AM, Gunanathan C, Potapov A, Efremenko I, Martin JML, Milstein D, Goldfarb D (2007) Gd^{3+} complexes as potential spin labels for high field pulsed EPR distance measurements. J Am Chem Soc 129(46):14138–14139. https://doi.org/10.1021/ja075544g
22. Potapov A, Song Y, Meade TJ, Goldfarb D, Astashkin AV, Raitsimring A (2010) Distance measurements in model bis-Gd(III) complexes with flexible "bridge". Emulation of biological molecules having flexible structure with Gd(III) labels attached. J Magn Reson 205(1):38–49
23. Berliner LJ (ed) (2002) Spin labeling, vol 14. Biological magnetic resonance. Springer US, New York. https://doi.org/10.1007/b114078
24. Bender C, Berliner LJ (eds) (2004) EPR: instrumental methods, vol 21. biological magnetic resonance, vol 21. Springer, New York. https://doi.org/10.1007/978-1-4419-8951-2
25. Hubbell WL, Altenbach C (1994) Site-directed spin labeling of membrane proteins. In: White SH (ed) Membrane protein structure: experimental approaches. methods in physiology. Springer, New York, pp 224–248. https://doi.org/10.1007/978-1-4614-7515-6
26. Hubbell WL, Gross A, Langen R, Lietzow MA (1998) Recent advances in site-directed spin labeling of proteins. Curr Opin Struc Biol 8(5):649–656
27. Berliner LJ, Grunwald J, Hankovszky HO, Hideg K (1982) A novel reversible thiol-specific spin label—papain active-site labeling and inhibition. Anal Biochem 119(2):450–455
28. Spaltenstein A, Robinson BH, Hopkins PB (1988) A rigid and nonperturbing probe for duplex DNA motion. J Am Chem Soc 110(4):1299–1301
29. Sale K, Song LK, Liu YS, Perozo E, Fajer P (2005) Explicit treatment of spin labels in modeling of distance constraints from dipolar EPR and DEER. J Am Chem Soc 127 (26):9334–9335
30. Anderson DJ, Hanson P, McNulty J, Millhauser G, Monaco V, Formaggio F, Crisma M, Toniolo C (1999) Solution structures of TOAC-labeled trichogin GA IV peptides from allowed ($g \approx 2$) and half-field electron spin resonance. J Am Chem Soc 121(29):6919–6927. https://doi.org/10.1021/ja984255c
31. Rassat A, Rey P (1967) Nitroxides. 23. preparation of amino-acid free radicals and their complex salts. B Soc Chim Fr 3:815–817
32. Seidemann R, Dulog L (1986) N-Carboxyanhydride of 4-Amino-4-Carboxy-2,2,6,6-Tetramethylpiperidine-1-Oxyl, a New Paramagnetic Monomer. Makromol Chem 187 (11):2545–2551
33. Toniolo C, Crisma M, Formaggio F (1998) TOAC, a nitroxide spin-labeled, achiral C(alpha)-tetrasubstituted alpha-amino acid, is an excellent tool in material science and biochemistry. Biopolymers 47(2):153–158
34. Martin L, Ivancich A, Vita C, Formaggio F, Toniolo C (2001) Solid-phase synthesis of peptides containing the spin-labeled 2,2,6,6-tetramethylpiperidine-1-oxyl-4-amino-4-carboxylic acid (TOAC). J Pept Res 58(5):424–432
35. Formaggio E, Broxterman QA, Toniolo C (2003) Synthesis of peptides based on C^{a}-tetrasubstituted α-amino acids. In: Goodman M, Felix A, Moroder L, Toniolo C (eds) Supplement—synthesis of peptides and peptidomimetics, vol E 22c, 4th edn. Houben-weyl methods of organic chemistry. Thieme, Stuttgart, pp 292–310

36. Salikhov KM, Tsvetkov YD (1979) Electron spin echo studies of spin-spin interactions in solids. In: Kevan L, Schwartz RN (eds) Time domain electron spin resonance. Wiley, New York, pp 231–278
37. Tsvetkov YD, Dzuba SA (1990) Pulsed ESR and molecular motions. Appl Magn Reson 1 (2):179–194
38. Rajca A, Kathirvelu V, Roy SK, Pink M, Rajca S, Sarkar S, Eaton SS, Eaton GR (2010) A spirocyclohexyl nitroxide amino acid spin label for pulsed EPR spectroscopy distance measurements. Chem-Eur J 16(19):5778–5782
39. Meyer V, Swanson MA, Clouston LJ, Boratynski PJ, Stein RA, McHaourab HS, Rajca A, Eaton SS, Eaton GR (2015) Room-temperature distance measurements of immobilized spin-labeled protein by DEER/PELDOR. Biophys J 108(5):1213–1219. https://doi.org/10.1016/j.bpj.2015.01.015
40. Shevelev GY, Gulyak EL, Lomzov AA, Kuzhelev AA, Krumkacheva OA, Kupryushkin MS, Tormyshev VM, Fedin MV, Bagryanskaya EG, Pyshnyi DV (2018) A versatile approach to attachment of triarylmethyl labels to DNA for nanoscale structural EPR studies at physiological temperatures. J Phys Chem B 122(1):137–143. https://doi.org/10.1021/acs.jpcb.7b10689
41. Fielding AJ, Concilio MG, Heaven G, Hollas MA (2014) New developments in spin labels for pulsed dipolar EPR. Molecules 19(10):16998–17025
42. Rossi R, Carpita A, Bellina F (1995) Palladium-mediated and/or copper-mediated cross-coupling reactions between 1-alkynes and vinyl, aryl, 1-alkynyl, 1,2-propadienyl, propargyl and allylic halides or related-compounds—a review. Org Prep Proced Int 27 (2):127–160
43. Chinchilla R, Najera C (2007) The sonogashira reaction: A booming methodology in synthetic organic chemistry. Chem Rev 107(3):874–922
44. Piton N, Schiemann O, Mu Y, Stock G, Prisner T, Engels JW (2005) Synthesis of spin-labeled RNAs for long range distance measurements by peldor. Nucleosides, Nucleotides Nucleic Acids 24(5–7):771–775
45. Gannett PM, Darian E, Powell JH, Johnson EM (2001) A short procedure for synthesis of 4-ethynyl-2,2,6,6-tetramethyl-3,4-dehydro-piperidine-1-oxyl nitroxide. Synthetic Commun 31(14):2137–2141
46. Gannett PM, Darian E, Powell J, Johnson EM, Mundoma C, Greenbaum NL, Ramsey CM Dalal NS, Budil DE (2002) Probing triplex formation by EPR spectroscopy using a newly synthesized spin label for oligonucleotides. Nucleic Acids Res 30(23):5328–5337
47. Frolow O, Bode BE, Angels JW (2007) The synthesis of EPR differentiable spinlabels and their coupling to uridine. Nucleos Nucleot Nucl 26(6–7):655–659
48. Piton N, Mu YG, Stock G, Prisner TF, Schiemann O, Engels JW (2007) Base-specific spin-labeling of RNA for structure determination. Nucleic Acids Res 35(9):3128–3143
49. Kolb HC, Finn MG, Sharpless KB (2001) Click chemistry: diverse chemical function from a few good reactions. Angew Chem Int Edit 40(11):2004–2021
50. Ding P, Wunnicke D, Steinhoff HJ, Seela F (2010) Site-directed spin-labeling of DNA by the Azide-Alkyne 'Click' reaction: nanometer distance measurements on 7-Deaza-2'-deoxyadenosine and 2'-deoxyuridine nitroxide conjugates spatially separated or linked to a 'dA-dT' base pair. Chem-Eur J 16(48):14385–14396
51. Jakobsen U, Shelke SA, Vogel S, Sigurdsson ST (2010) Site-directed spin-labeling of nucleic acids by click chemistry: detection of abasic sites in duplex DNA by EPR spectroscopy. J Am Chem Soc 132(30):10424–10428
52. Barhate N, Cekan P, Massey AP, Sigurdsson ST (2007) A nucleoside that contains a rigid nitroxide spin label: a fluorophore in disguise. Angew Chem Int Edit 46(15):2655–2658
53. Marko A, Denysenkov V, Margraft D, Cekan P, Schiemann O, Sigurdsson ST, Prisner TF (2011) Conformational flexibility of DNA. J Am Chem Soc 133(34):13375–13379
54. Tkach I, Pornsuwan S, Hobartner C, Wachowius F, Sigurdsson ST, Baranova TY, Diederichsen U, Sicoli G, Bennati M (2013) Orientation selection in distance measurements

between nitroxide spin labels at 94 GHz EPR with variable dual frequency irradiation. Phys Chem Chem Phys 15(10):3433–3437
55. Stoller S, Sicoli G, Baranova TY, Bennati M, Diederichsen U (2011) TOPP: a novel nitroxide-labeled amino acid for EPR distance measurements. Angew Chem Int Edit 50 (41):9743–9746
56. Fedin MV, Shevelev GY, Pyshnyi DV, Tormyshev VM, Jeschke G, Yulikov M, Bagryanskaya EG (2016) Interaction of triarylmethyl radicals with DNA termini revealed by orientation-selective W-band double electron-electron resonance spectroscopy. Phys Chem Chem Phys 18(42):29549–29554. https://doi.org/10.1039/c6cp05904j
57. Savitsky A, Dubinskii AA, Flores M, Lubitz W, Mobius K (2007) Orientation-resolving pulsed electron dipolar high-field EPR spectroscopy on disordered solids: I. Structure of spin-correlated radical pairs in bacterial photosynthetic reaction centers. J Phys Chem B 111 (22):6245–6262. https://doi.org/10.1021/jp070016c
58. Denysenkov VP, Prisner TF, Stubbe J, Bennati M (2006) High-field pulsed electron-electron double resonance spectroscopy to determine the orientation of the tyrosyl radicals in ribonucleotide reductase. P Natl Acad Sci USA 103(36):13386–13390. https://doi.org/10.1073/pnas.0605851103
59. Denysenkov VP, Biglino D, Lubitz W, Prisner TF, Bennati M (2008) Structure of the tyrosyl biradical in mouse R2 ribonucleotide reductase from high-field PELDOR. Angew Chem Int Edit 47(7):1224–1227
60. van Amsterdam IMC, Ubbink M, Canters GW, Huber M (2003) Measurement of a Cu-Cu distance of 26 Å by a pulsed EPR method. Angew Chem Int Edit 42(1):62–64
61. Becker JS, Saxena S (2005) Double quantum coherence electron spin resonance on coupled Cu(II)-Cu(II) electron spins. Chem Phys Lett 414(1–3):248–252
62. Elsasser C, Brecht M, Bittl R (2002) Pulsed electron-electron double resonance on multinuclear metal clusters: assignment of spin projection factors based on the dipolar interaction. J Am Chem Soc 124(42):12606–12611
63. Roessler MM, King MS, Robinson AJ, Armstrong FA, Harmer J, Hirst J (2010) Direct assignment of EPR spectra to structurally defined iron-sulfur clusters in complex I by double electron-electron resonance. P Natl Acad Sci USA 107(5):1930–1935
64. Garbuio L, Bordignon E, Brooks EK, Hubbell WL, Jeschke G, Yulikov M (2013) Orthogonal spin labeling and Gd(III)-nitroxide distance measurements on bacteriophage T4-lysozyme. J Phys Chem B 117(11):3145–3153
65. Potapov A, Yagi H, Huber T, Jergic S, Dixon NE, Otting G, Goldfarb D (2010) Nanometer-Scale distance measurements in proteins using Gd^{3+} spin labeling. J Am Chem Soc 132(26):9040–9048
66. Banerjee D, Yagi H, Huber T, Otting G, Goldfarb D (2012) Nanometer-range distance measurement in a protein using Mn^{2+} tags. J Phys Chem Lett 3(2):157–160
67. Bagryanskaya EG, Krumkacheva OA, Fedin MV, Marque SRA (2015) Development and application of spin traps, spin probes, and spin labels. Method Enzymol 563:365–396
68. Goldfarb D (2014) Gd^{3+} spin labeling for distance measurements by pulse EPR spectroscopy. Phys Chem Chem Phys 16(21):9685–9699. https://doi.org/10.1039/c3cp53822b
69. Akhmetzyanov D, Schop P, Marco A, Kunjir NC, Sigurdsson ST, Prisner TF (2013) Phys Chem Chem Phys 15:619–627

Chapter 5
Spatial Distribution of Nitroxyl Radicals

Abstract The nitroxylfree radicals, also known as nitroxide or aminoxyl radicals, are convenient, spectroscopically-active probes for many physicochemical studies. They have long been used to study the spatial distribution of paramagnetic centers in solids using PELDOR or DEER spectroscopy. The spatial distribution function $F(r)$ between radicals can be determined from their distance-dependent dipolar interactions. The detailed spatial distribution of reactive species, particularly radicals or ion-radicals, often determines the kinetics of their reactions in solids. Nitroxyl radicals have been widely used for pulsed EPR experimental studies of magnetic relaxation processes, such as spectral and instantaneous diffusion, cross-relaxation, and spin-lattice relaxation. Their well-established chemical structure and stability make nitroxyls convenient, spectroscopically-active probes for many physicochemical studies. One area in which nitroxyls have been employed concerns the spatial distribution of paramagnetic centers in solids. The detailed spatial distribution of reactive species, particularly radicals or ion-radicals, often determines the kinetics of their reactions in solids. The spatial distribution function $F(r)$ between radicals can be determined from their distance-dependent dipolar interactions using, e.g., ESE (Salikhov et al in Electron spin echo and its applications. Nauka, Novosibirsk, 1976 [1]; Salikhov and Tsvetkov in Time domain electron spin resonance. Wiley Interscience, New York, pp 232–277, 1977 [2]) or PELDOR techniques. In this chapter, we describe a few examples of the variety of spatial distributions that have been examined by PELDOR methods. Studies at X-band using nitroxyls and nitroxyl-labeled compounds were chosen to emphasize the distance distributions and to minimize complications from changes in mw frequency or spin properties at high frequency.

5.1 Uniform Distribution in 2- and 3-Dimensions

The literature is replete with pulsed EPR studies of spins distributed randomly and uniformly in a volume or on a surface. Many of these studies were made in frozen solutions where formation of a glass helps to avoid heterogeneity in the distribution of the solutes, particularly the spins. Solutions that freeze into a polycrystalline

© Springer Nature Switzerland AG 2019

Y. D. Tsvetkov et al., *Pulsed Electron–Electron Double Resonance*,
https://doi.org/10.1007/978-3-030-05372-7_5

solid often have very inhomogeneous distributions of radicals. The solvent freezes into rather pure crystallites with the radicals expelled and concentrated into the interfaces and interstices between crystallites: essentially a separation of phases into relatively pure solvent crystals and very concentrated solute.

For a random, uniform volume distribution of spins in a solid, the PELDOR time trace $V(T)$ decays by the exponential law with $d = 3$ in Eq. 1.18. A representative study was made of a uniform distribution of the nitroxyl radical often called tempone, **5–1**, Table 5.1. Where convenient, we will refer to the radicals by number because the chemical name is too cumbersome while many radicals have multiple acronyms that are not universally recognized and not unique.

The $V(T)$ for **5–1** in frozen water-glycerol solutions were obtained over a wide concentration range at 77 K [3]. The decay of $V(T)$ is exponential over the full range of T with the expected dimensionality $d = 3$, Fig. 5.1. The linear decay in a semilog plot is characteristic of a random 3D distribution of spins.

A planar spatial distribution was found for spin-labeled peptides adsorbed on spherical nanoparticles [4] and for the complex of a labeled protein on a membrane [5]. In these cases, the experimental dimensionality d was within 10% of the theoretical $d = 2$ value. As an example, data is shown for a mono spin-labeled peptide trichogen **5–4** adsorbed on the surface of SiO_2 nanoparticles, Fig. 5.2. In this case, $V(T)$ decays with $d/3 = 0.681$, i.e., a dimensionality of $d = 2.043$. On a

Table 5.1 Structures and common abbreviations of radicals; TOAC indicates **4-22**

	Radical	Common Names
5–1	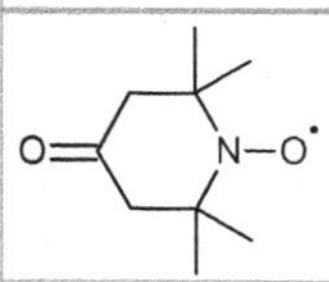	Tempone, Tanone
5–2	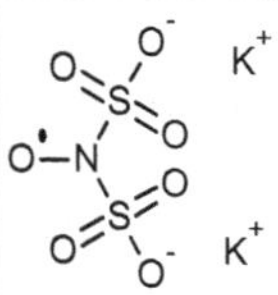	Fremy's Salt, nitrosodisulfonate, peroxylamine disulfonate
5–3	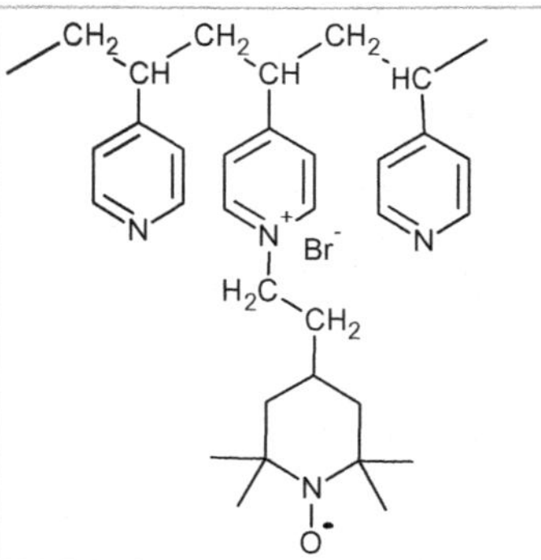	P-4-VP
5–4	n-Oct-Aib-Gly-Leu-TOAC-Gly-Gly-Leu-Aib-Gly-Ile-Leu-OMe	O-Tri[4]

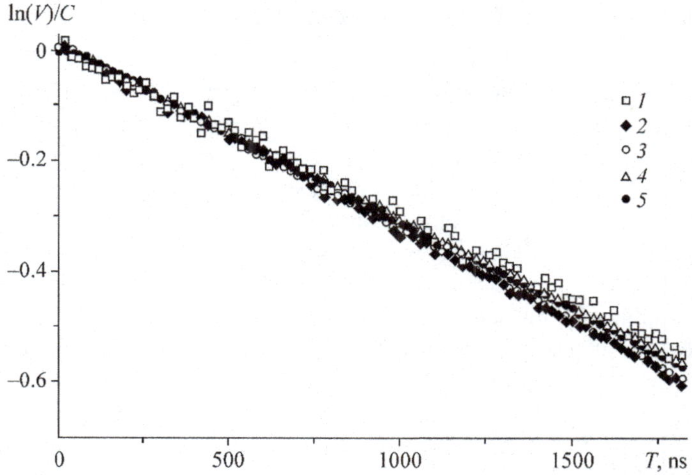

Fig. 5.1 Experimental PELDOR time traces for **5–1** in frozen aqueous glycerol solutions (30% H_2O) at radical concentrations: *1* 1.7×10^{17} cm^{-3}; *2* 4×10^{17}; *3* 6×10^{17}; *4* 9×10^{17}; and *5* 1.4×10^{18}; the logarithm of the time trace is divided by the **5–1** concentration, so that the superimposed, linear dependences reveal first-order decays with rates proportional to **5–1** concentration. Reprinted from Milov and Tsvetkov [3], with permission of Springer, copyright 2000

semilog plot such as Fig. 5.1, $V(T)$ is a straight line only for $d = 3$; the curvature indicates the dimensionality of the spatial distribution.

In these examples, the solution, surface or membrane containing the nitroxyl radical controlled the dimensionality $d = 2$ or 3. When spin-labeled peptides were adsorbed on organic particles whose surface had a fractal dimensionality, the $V(T)$ reflect that fractal nature [6]. See Sect. 6.1.4 for additional discussion of peptides on surfaces.

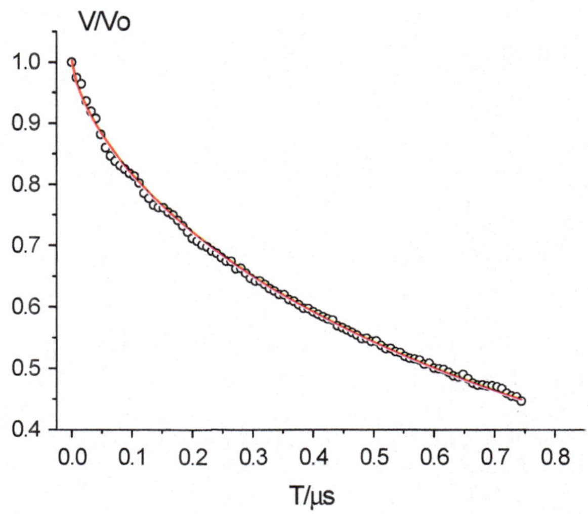

Fig. 5.2 PELDOR time trace for mono spin-labeled trichogin **5–4** adsorbed on the surface of SiO_2 nanoparticles: the *red line* shows the normalized decay function $V/V_0 = \exp(-0.982T^{d/3})$ with $d/3 = 0.681$ (Milov et al. Unpublished)

5.2 Electrostatic Effects on Spatial Distribution

Deviations from a random spin distribution are reflected in the decay of the PELDOR time trace. For example, when the local spin concentration exceeds the average value, e.g., in spin pairs, radical clusters, or radiation tracks, the $V(T)$ decreases quite rapidly at small times T because of the nearby spins and their strong dipolar interactions. On the other hand, the opposite occurs when the close approach of spins is hindered for some reason. Fewer nearby spins result in fewer strong dipole interactions. Hence, $V(T)$ decays less at short times than for a random spin distribution with the same average concentration.

Electrostatic effects, for instance, hinder or encourage the close approach of charged particles or ion-radicals. Such an electrostatic effect is seen by 3pPELDOR for the simplest nitroxyl ion-radical, known as Fremy's salt, **5–2**, at 77 K [3].

The $V(T)$ for **5–2** in glassy, frozen aqueous glycerin solutions, Fig. 5.3, are qualitatively different from those of the uncharged **5–1**, Fig. 5.1. The time traces of **5–2** do not fit the general decay law for a random radical distribution. The plot of $\ln(V)/C$ versus T has deviations from linearity which are most pronounced at low concentrations of **5–2**, Fig. 5.3. The curves are not superimposed, as they were for uncharged **5–1**, and the initial decay rate at small $T \sim 100$ ns is considerably slower

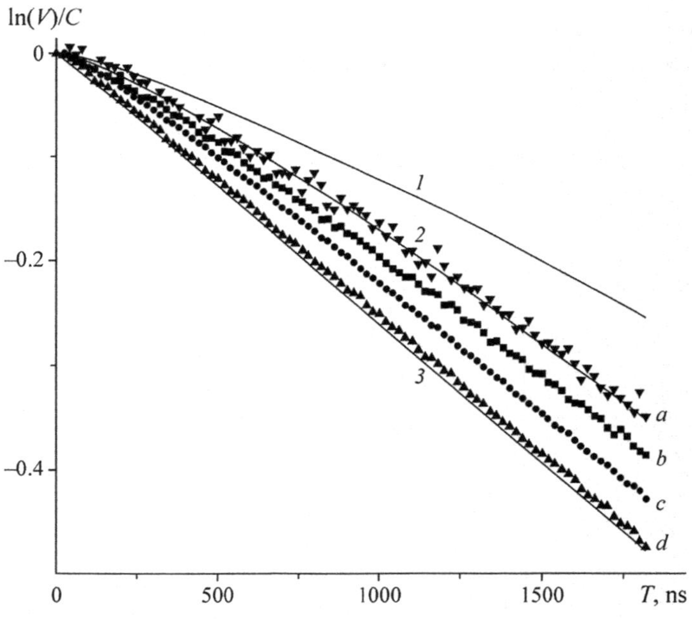

Fig. 5.3 Experimental and calculated PELDOR time traces normalized to the concentration of **5–2** in aqueous glycerin solution at 77 K: *Points* denote the experimental values at the following concentrations: *a* 2.6×10^{17} cm^{-3}; *b* 8×10^{17}; *c* 2.2×10^{18}; and *d* 8×10^{17} in 0.3 M K_2CO_3. *Solid lines* show calculations for radical concentrations of 2.6×10^{17} cm^{-3} at different ionic strengths: *1* 0 M, *2* 1.3×10^{-3}, *3* 1.0. Reprinted from Milov and Tsvetkov [3], with permission of Springer, copyright 2000

than that for uniformly distributed radicals. However, the decay rates become faster as T increases, until the normalized traces become parallel beyond 1000 ns. This means that at long T the decay rates become proportional to radical concentration, as they were for the uncharged **5–1**.

The initial decay rate is increased when the ionic strength of the solution is increased by addition of K_2CO_3, see traces *b* and *d*, and $ln\,(V)/C$ becomes linear over its full length. The slower decay of $ln\,(V)/C$ at short T for **5–2** indicates that the fraction of nearby spins with fairly strong dipole-dipole coupling is less. Increasing the ionic strength by addition of K_2CO_3 allows the added K^+ and $CO_3{}^{2-}$ ions to screen the electrostatic repulsion between ion-radicals. As a result, the **5–2** ion radicals become more randomly distributed. A similar screening effect occurs as the **5–2** concentration is increased in the absence of K_2CO_3, because the **5–2** ions also contribute to the ionic strength of the sample.

The effect of mutual repulsion between radical ions on the phase relaxation in ESE and PELDOR traces can be quantitatively described in dilute solutions by classical Debye-Hückel theory [7]. The concentration $C(r)$ of ion radicals at a distance r from an ion radical is

$$C(r) = Ce^{-\frac{Q}{r}e^{-\chi r}} \tag{5.1}$$

$$Q = \frac{Z^2\varepsilon^2}{D\,kT} \tag{5.2}$$

$$\chi^2 = 8\pi\varepsilon^2 \frac{\boldsymbol{I}}{D\,k_B T} \tag{5.3}$$

$$\boldsymbol{I} = 1/2\sum_i C_i Z_i^2 \tag{5.4}$$

where C is the mean concentration of ion radicals and Z is their charge in units of the electron charge, ε; D is the dielectric constant; k_B is the Boltzmann constant; T is temperature in Kelvin; χ is the Debye-Hückel constant; and $\boldsymbol{I}$ is the ionic strength of the solution with Z_i and C_i the charge and concentration, respectively, of the i-th type of ions.

The distance distribution of B spins with respect to A spins is described by Eq. (5.1) for the **5–2** ion radical. This makes it possible to calculate the PELDOR time trace $V(T)$ and spin concentration $C(r)$ with the general expression for $V(T)$ Eq. 1.15 by the Markov method. Such calculations describe the experimental data quite closely, Fig. 5.3. Curve 1 describes the decay in the limit of zero ionic strength and indicates the maximum possible influence of charge on $V(T)$ relaxation kinetics of this system.

At a given ionic strength, Q and χ determine the scale of distances over which the approach of ion radicals is hindered. Taking r_0 as the distance at which $C(r_0) = C/2$, we find $r_0 = 3.0$ nm at the lowest **5–2** concentration of 2.6×10^{17} cm^{-3}. This distance is a few nm larger than the physical radius of **5–2** and affects the initial PELDOR time trace for $T \lesssim r_0^3/(\gamma^2\hbar)$ in the microsecond region.

Debye-Hückel theory provides a satisfactory description of the spatial distribution of ion radicals in frozen glassy solutions and allows the quantitative description of $V(T)$ and magnetic phase relaxation.

5.3 Spin Labels on a Polymer

5.3.1 Uniform Linear Distribution

CW EPR spectroscopy of spin-labeled polymers is widely used for studying the conformation and structure of polymer molecules in the solid phase [8]. The inhomogeneously-broadened EPR spectra of commonly-used nitroxyl labels have a total width of several tens of Gauss in the solid phase. Weak dipole-dipole interactions between labels are hidden within the lineshape, severely limiting the range of dipole-dipole interactions that can be studied by CW EPR. The PELDOR technique opened up new possibilities to measure dipolar interactions independent of inhomogeneous broadening. We examine the application of PELDOR to the spatial distribution of nitroxyl spin labels attached to an organic polymer chain [9] in detail here.

The PELDOR technique was used to study phase relaxation due to spin-spin interactions in spin-labeled poly-4-vinylpyridine, **5–3** [9]. Glassy frozen solutions of **5–3** in deuterated glycerin and methanol at 77 K were used to study weak spin-spin interactions at large distances not accessible to CW EPR methods. The **5–3** material had been studied by CW EPR and had an average molecular weight of 50,000 Dalton and a degree of polymerization $N = 476$ [9]. Polymers with two degrees of labeling ($\beta = 0.06$ and 0.09) were used, where β is the ratio of the number of labels to the degree of polymerization N, i.e., labels per monomer unit. Solutions of **5–3** in deuterated glycerin were prepared by dissolving polymer in deuterated methanol and adding it to glycerin to the desired polymer concentration. Solutions of **5–1** radicals in the same solvents were prepared similarly as a reference. The mean concentration of spin labels was confirmed by CW EPR.

The PELDOR time traces $V(T)$ of **5–3** with $\beta = 0.09$ show an initial rapid decay followed by a slower decay that depends on polymer concentration, Fig. 5.4, curves 2 and 3. Similar $V(T)$ were observed for samples with $\beta = 0.06$, and for solutions of **5–3** in methanol. In contrast, frozen solutions of **5–1** nitroxyls in the same solvents have the expected exponential decay of radicals randomly distributed in three dimensions, Fig. 5.4, curve *1*.

The *intra*molecular contribution $V_{INTRA}(T)$ from interactions between spins in a single polymer molecule was determined by extrapolation of $V(T)$ to zero concentration with constant mw pump pulse parameter p_B. The $ln\,(V)$ at different values of T, Fig. 5.5, depend linearly on spin label concentration as

$$ln\,(V) = ln\,(V_0) - AC \tag{5.5}$$

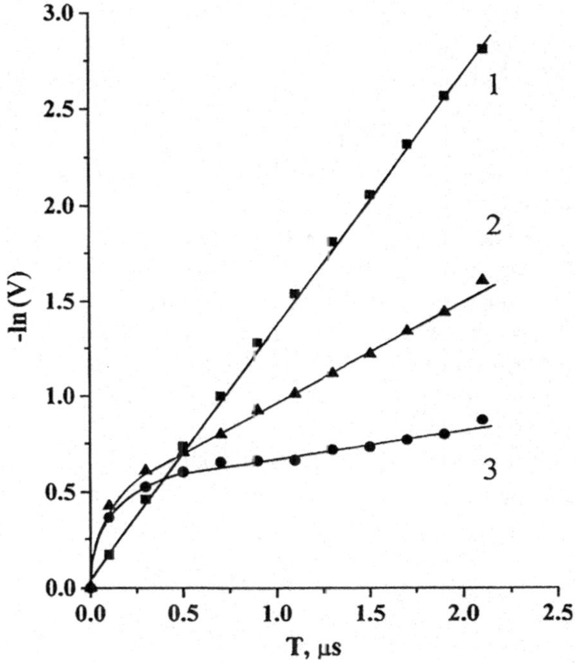

Fig. 5.4 PELDOR amplitude as a function of time T: *1* for **5–1** at a concentration of 3.6×10^{18} cm^{-3}; *2* and *3* solutions of **5–3** in glycerin with $\beta = 0.09$ at a mean concentration of labels of 6.4×10^{18} cm^{-3} and 1×10^{18} cm^{-3}, respectively. Reprinted from Milov and Tsvetkov [9], with permission of Springer, copyright 1997

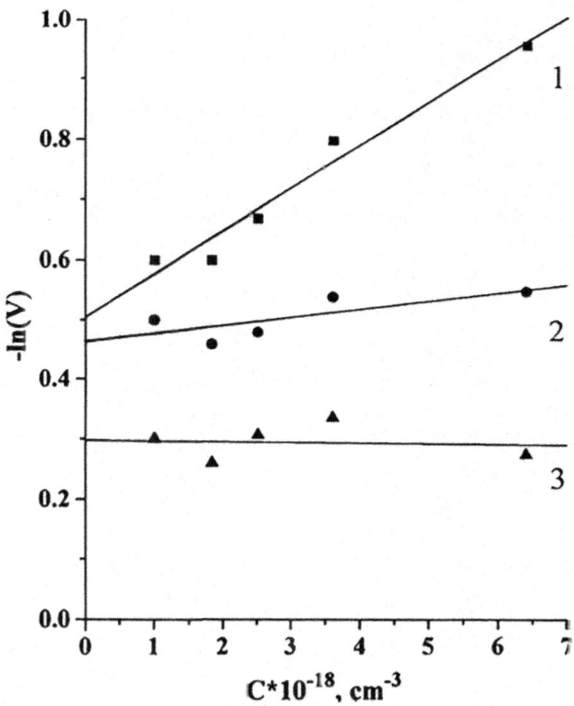

Fig. 5.5 PELDOR amplitude as a function of the mean spin-label concentration for **5–3** with $\beta = 0.09$ in glycerin at 77 K; the values of T for curves *1*, *2*, and *3* are 2.0, 1.0 and 0.3 μs, respectively. Reprinted from Milov and Tsvetkov [9], with permission of Springer, copyright 1997

where C is the mean concentration of spin labels in the sample; and A and $ln\,(V_0)$ are the experimental slope and intercept. The AC term in Eq. (5.5) is the *inter*molecular spin-spin term, $ln\,(V_{INTER}(T))$, while $ln\,(V_0)$ is phase relaxation from spin labels within each polymeric molecule, $ln\,(V_{INTRA}(T))$. A linear relation of identical form holds for **5–3** in methanol.

A plot of $ln(ln|V_0|)$ reveals a linear dependence on $ln\,(T)$, Fig. 5.6. Thus, the phase relaxation due to intramolecular spin label interactions is given by Eq. 1.18, with dimensionality $d = 1.0 \pm 0.1$, determined from the slopes of $\sim -1/3$ in Fig. 5.6.

The intramolecular spin-spin interactions show fractal effects in the spatial distribution of spins within each macromolecule. Therefore, Eqs. 1.19–1.22 were used to determine the pair distribution function for labels along the polymer chain. The experimental dimensionality d is close to unity and indicates a linear spatial distribution. The one-dimensional concentration C_1 of spin labels is 1.7×10^6 cm^{-1} for $\beta = 0.06$ and 2.6×10^6 cm^{-1} for $\beta = 0.09$. As expected, C_1 is proportional to the degree of polymer labeling, making it possible to represent the pair distribution function as:

$$F(r) = \frac{C_1}{2\pi r^2} = 4.5 \times 10^6 \beta r^{-2} \text{cm}^{-3} \qquad (5.6)$$

The errors in C_1 are due to the uncertainty of about 30% in the absolute p_B values.

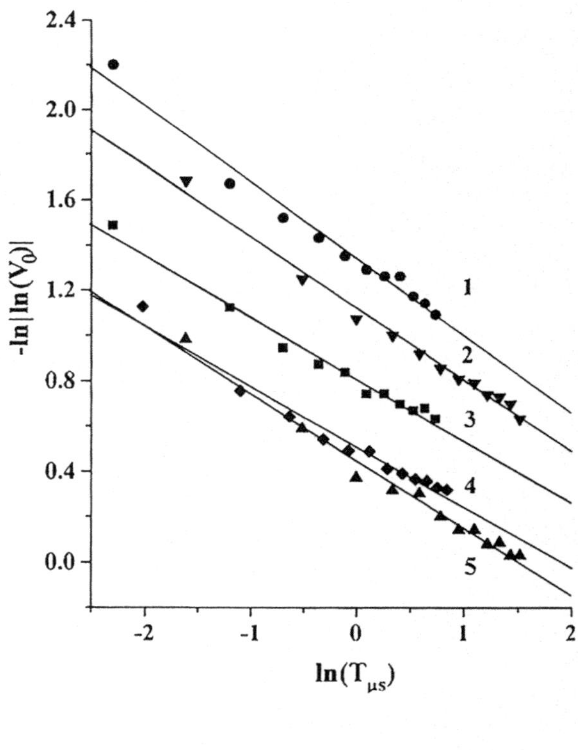

Fig. 5.6 Dependence of the echo signal V_0 on time T extrapolated to zero spin label concentration: *1*, *2*, *3*, and *5* refer to **5–3** in glycerin, with: *1* $\beta = 0.09$, $p_B = 0.08$; *2* $\beta = 0.06$, $p_B = 0.14$; *3* $\beta = 0.09$, $p_B = 0.14$; *5* $\beta = 0.06$, $p_B = 0.29$, respectively; *4* refers to **5–3** in methanol, with $\beta = 0.06$, $p_B = 0.24$. Reprinted from Milov and Tsvetkov [9], with permission of Springer, copyright 1997

It is important to note that the $V_{INTER}(T)$ decay also is not a simple exponential law even though the polymer molecules are randomly distributed in the frozen solution. The slope A in Eq. (5.5) characterizes intermolecular spin-spin interactions and is given as $A/p_B = (8\pi^2/9\sqrt{3})\gamma^2\hbar T$ through Eq. 1.17 for a random, three-dimensional distribution of spins. Yet, the experimental ratio A/p_B for **5–3** in glycerin with $\beta = 0.06$ or 0.09 is clearly not linear in T, Fig. 5.7.

The decay due to intermolecular spin-spin interactions for labels on a polymer is substantially less efficient than that for randomly distributed spins. The experimental A/p_B values are described much better by the parabola $A/p_B = 7.36 \times 10^{-8}$ T^2 cm^3 s^{-2}, Fig. 5.7, curve *3*.

This result seems rather unexpected for intermolecular spin-spin interactions. The quadratic dependence on T indicates that strong dipolar interactions between labels of different macromolecules are absent, i.e., the spin labels do not approach each other very closely. To estimate the range of closest approach between spin labels, the A/p_B was calculated for a model where no B spins are located inside a sphere of radius R; but, at distances longer than R, the B spins are randomly distributed. The A/p_B, calculated at various R, are shown by solid curves *2*, *3*, and *4*, Fig. 5.7. At $R = 0$ nm, we obtain a linear dependence from the usual exponential law for a random distribution of spins in space (the *dots* in line *1*). Curves *2*, *3*, and *4* were obtained for $R = 9.0$, 13.5, and 18.0 nm, respectively. Curve *3*, with $R = 13.5$ nm, describes the

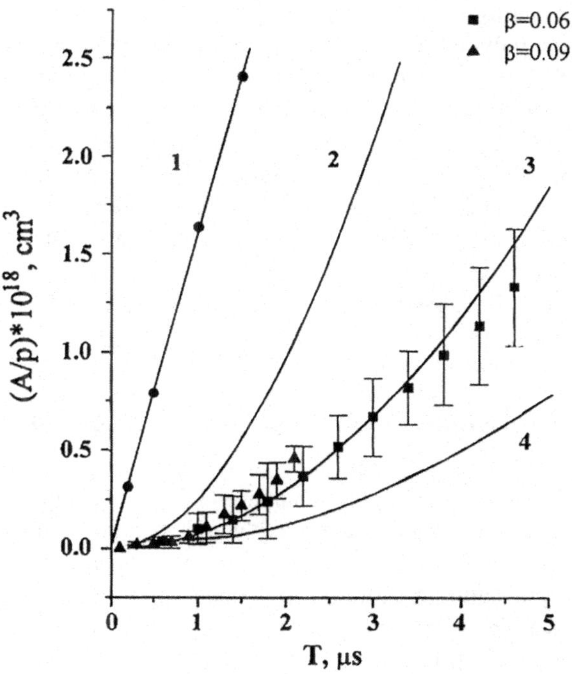

Fig. 5.7 Calculated and experimental dependence of the inter-macromolecular phase relaxation efficiency A/p_B on T for **5–3** in glycerin; *Line 1* was obtained for a random spin distribution with $d = 3$ and the *dots* on this line were calculated for $R = 0$, Curve *2*, *3*, and *4* were calculated for an excluded volume of radius R= 9.0, 13.5 and 18.0 nm, respectively, around each polymer. Reprinted from Milov and Tsvetkov [9], with permission of Springer, copyright 1997

experimental data rather well. This provides an estimation of the range of closest approach for labels on different macromolecules.

The results for intra- and intermolecular interactions between spin labels can be unambiguously explained by electrostatic repulsion between the charged fragments of the polymeric chain. This repulsion leads to the extension of each macromolecule into a linear chain and prevents different macromolecules, and their labels, from approaching each other closer than 10.0 nm. The charges occur at the spin-labeled sites of **5–3** and where the nitrogens of some of the pyridine groups of **5–3** happen to be protonated.

The conclusion that **5–3** has an extended conformation differs from CW EPR results showing that the conformation of **5–3** in glassy methanol is a Gaussian knot [10, 11]. This difference probably is because the strongest interactions between spin labels at the shortest distances dominate the broadening of the EPR spectra while PELDOR is most sensitive to larger distances. The length and flexibility of the polymer side chain connecting spin labels to the main polymer chain may produce subtle, but significant, distortions in $C(r)$ at the short distances r probed by the CW EPR methods. The actual $C(r)$ for short distances may be closer to that of a Gaussian knot than to an idealized linear model. In addition, the charging of the polymer and, as a result, its conformation can depend strongly on solvent pH in the different samples.

The linear spatial distribution seems to be characteristic of highly-charged polymers. A PELDOR study of **5–2** anions associated with an unlabeled polyelectrolyte, i.e., highly-charged polymer, in frozen water-glycerol solutions found the polymer in an extended conformation that persisted for at least 5 nm [12]. The **5–2** anions were randomly distributed along the extended, linear polyelectrolyte, producing a PELDOR time trace decay with $d = 1$. In another study, dipolar modulation of biradicals composed of nitroxyl labels attached to the ends of chains of CH_2-groups was measured [13]. The modulation showed extended, almost completely stretched, chains for the entire range of biradicals with 8–20 CH_2-groups **3.1.1**. The biradicals were uncharged and force field calculations predict extended chains, so electrostatic interactions are not the only factor favoring extended, linear polymers. These last two studies are consistent with those on **5–3**.

5.4 Spin-Labeled Peptides in Bacterial Cells

The antibiotic peptide trichogin GA IV and spin-labeled derivatives, such as **5–4**, have been studied extensively in the outer membrane of spherical cells of the Gram-positive bacterium *Micrococcus luteus* [14]. The rather simple membrane architecture of Gram-positive microorganisms is an advantage in such studies. The membrane is composed of a single, lipid bilayer surrounded by a thick peptidoglycan cell wall.

Cells of *M. luteus* have a maximum radius of 500 nm. Singly spin-labeled trichogin **5–4** was introduced into the cells by a special procedure [14], which

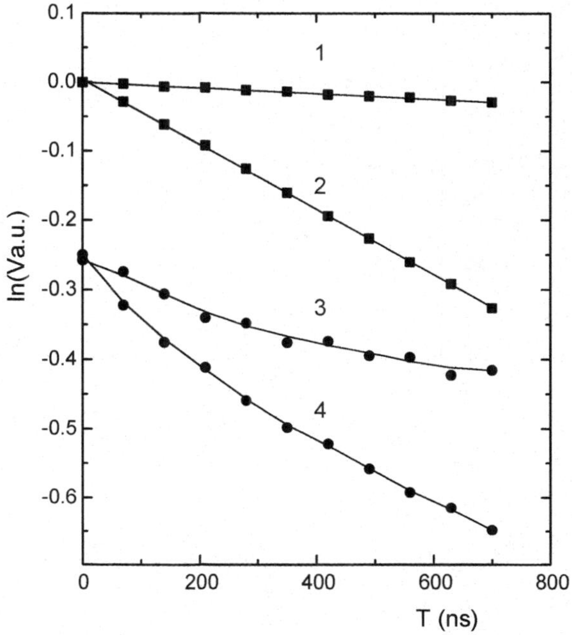

Fig. 5.8 A plot of the experimental *ln* (V) versus T for **5–4** at 77 K: *1* and *2*—frozen glassy solutions in ethanol with spin label concentrations of 2.7×10^{-4} and 2.7×10^{-3} M, respectively; *3* and *4* frozen suspensions in phosphate buffer of **5–4** bound to *M. luteus* cells with mean spin label concentrations of 1.1×10^{-4} and 2.8×10^{-4} M, respectively. Reprinted from Milov et al. [14], with permission of Springer, copyright 2002

produced a dry pellet of cells for EPR and PELDOR measurements at 77 K. Frozen, glassy ethanol solutions of the same mono-labeled **5–4** were used as a reference.

PELDOR time traces $V(T)$ were obtained for reference samples, curves *1* and *2*, and cell pellets, curves *3* and *4*, Fig. 5.8. The $V(T)$ show an important difference. The logarithm of $V(T)$ for **5–4** in ethanol decreases linearly with T, indicating an exponential decay of the signal. On the other hand, the cell data have significant curvature, indicating a more complicated decay. In addition, curves *1* and *4* have about the same mean concentration of **5–4**, yet the dipole-dipole relaxation is ten-fold more efficient in the cell sample than in ethanol solution. A rough estimation shows that the local concentration of **5–4** in the cell pellet is about 2.7×10^{-3} M, much larger than the average concentration. This means that the **5–4** is restricted to a small portion of the cell, comparable in size of the cell membrane which is about 7 nm thick including the membrane proteins bulging out of the lipid bilayer. Detailed analysis of these $V(T)$ provides rather precise information about the distribution of spin labels in the membrane. ESE and PELDOR data for **5–4** bound to *M. luteus* excludes the possibility of membrane-bound aggregates, so only non-aggregated **5–4** is present.

Replotting the experimental cell data as *ln* (*ln*$|(V)|$) versus *ln* (T), gives straight parallel lines, Fig. 5.9, described by the exponential law, Eq. 1.18 with a fractal dimensionality given by $d/3 = 0.71 \pm 0.012$ that is independent of the **5–4** concentration. This value is slightly larger than the $d/3 = 2/3$ for an ideal planar surface, yet the deviation is statistically significant and much larger than the

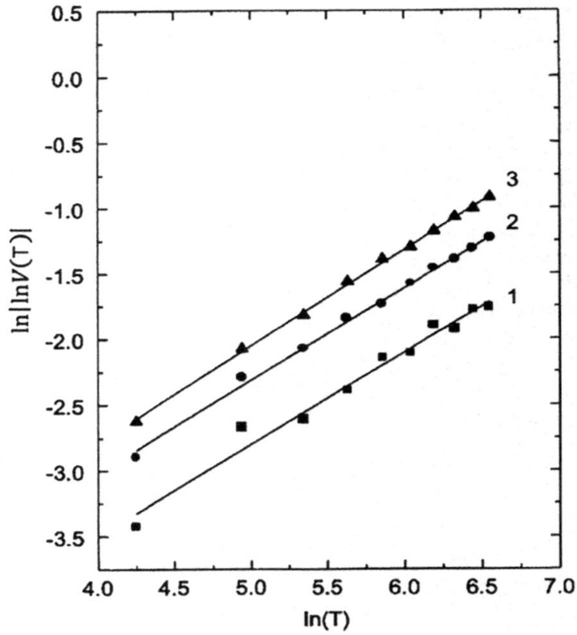

Fig. 5.9 Experimental $V(T)$: Curves *1-3*: **5–4** bound to the *M. luteus* cell at 77 K with mean concentrations of **5–4** of 1.1×10^{-4}, 2.0×10^{-4}, and 2.8×10^{-4} M, respectively. Reprinted from Milov et al. [14], with permission of Springer, copyright 2002

experimental uncertainty. There is an unfortunate typographical error that inflates the uncertainty for $d/3$ in the original paper when its Fig. 4 is discussed [14]. There are three ways that a deviation from $d/3 = 2/3$ could arise. One is a curvature of the cell membrane causing a slight deviation from planarity, but that has too small an effect for *M. luteus* cells with a radius of 500 nm, or even for cells with half that radius. The other two possibilities are that the labels are randomly distributed in the bulk of the membrane or at both membrane surfaces—basically, in a thin slab or on two parallel planes.

Monte-Carlo simulations of PELDOR time traces based on Eq. 1.16 were made for both these possibilities and the value of $d/3$ extracted. The cell membrane was modeled as a spherical layer with thickness h and inner radius R. Two different distributions of spin labels were considered: (i) labels in the volume of this layer and (ii) labels at both surfaces of the spherical layer [14]. In both cases, the spins were located independently of one another. The simulated $d/3$ values are plotted versus the membrane thickness h for $R = 500$ nm, Fig. 5.10. The experimental value of $d/3 = 0.71 \pm 0.012$ is indicated by the horizontal line.

Both simulations converge to the ideal value of $d/3 = 2/3$ when the membrane thickness h vanishes. Line *1* for the 'slab' distribution increases monotonically as h increases, approaching the limiting value of 3/3 for a three-dimensional random distribution. Line *1* intersects the experimental value at $h = 2.4 \pm 0.3$ nm suggesting that **5–4** lies in a layer of this thickness at the membrane. This could be at the center of the membrane or at the outer surface of either the membrane or the cell wall.

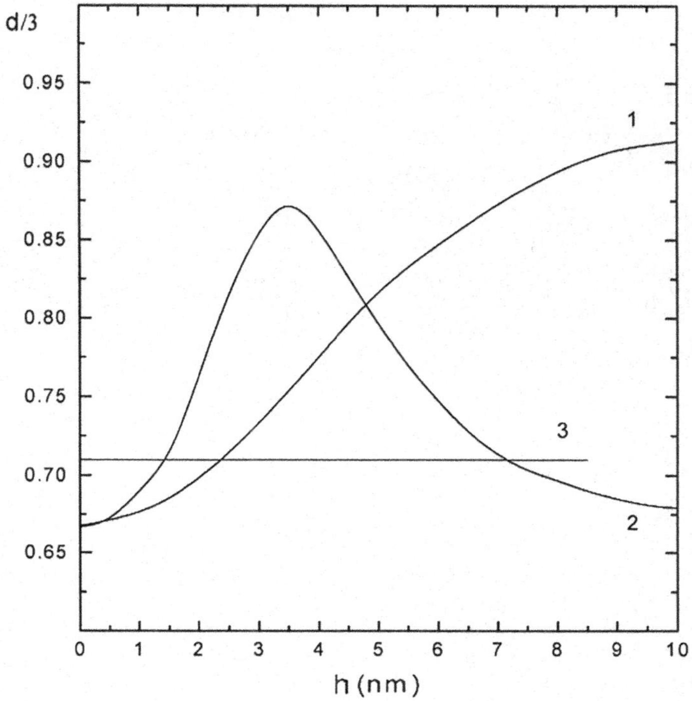

Fig. 5.10 Simulated dimensionality $d/3$ for a spherical layer with thickness h and radius R = 500 nm: *1* labels located in the bulk of the layer; *2* labels located on both surfaces of the layer; *3* experimental value $d/3$ = 0.71 ±0.012. Reprinted from Milov et al. [14], with permission of Springer, copyright 2002

The simulated curve *2* for the two-surface distribution crosses the experimental value twice, Fig. 5.10. The first crossing is at $h = 1.5 \pm 0.2$ nm, much smaller than the known membrane or cell wall thickness and does not seem realistic. The second crossing occurs at $h = 7.0 \pm 0.3$ nm, larger than the thickness of the lipid bilayer, but in good agreement with the currently accepted thickness of 7.5 nm for the lipid/ protein membrane of *M. luteus* cells [15]. This would place **5–4** at both surfaces, external and internal, of the membrane; possibly bound to proteins bulging out of the lipid bilayer.

At this time, neither possibility for the location of trichogin can be ruled out [14]. Trichogin could be distributed randomly on both the inner and the outer surfaces of the membrane with a distance of 7 nm between the surfaces; or trichogin could be distributed randomly in a ~2.4 nm thick layer at the outer membrane surface, Fig. 5.11. The main target of trichogin, after penetrating the porous cell wall, is the cytosolic membrane. In any case, it is not clear whether trichogin binds to phospholipids, whose major constituents are cardiolipin and phosphatidyl glycerol [16, 17], or to membrane proteins, which comprise 49% *w/w* of the membrane [18].

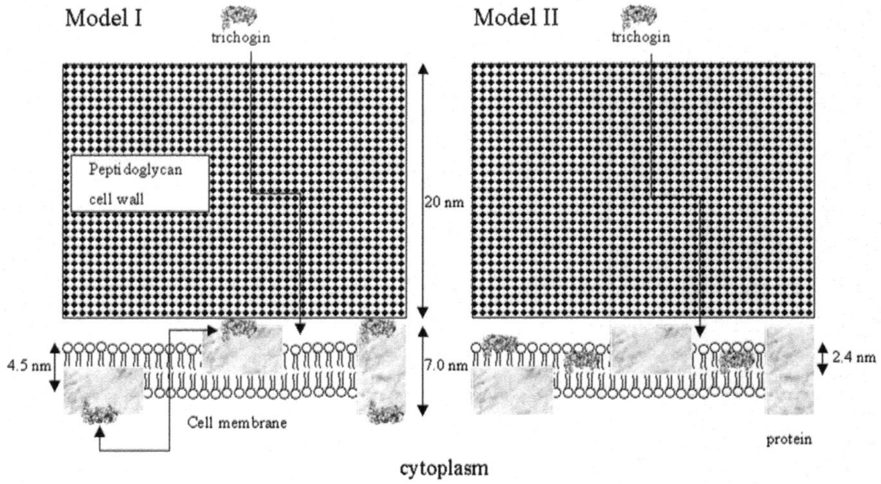

Fig. 5.11 Two proposed binding modes of trichogin GA IV to *M. luteus* cells: *I* bound to membrane proteins bulging out of both sides of the lipid bilayer; or *II* bound at the periplasmic side of the membrane and distributed throughout a layer of 2.4 nm thickness. Reprinted from Milov et al. [14], with permission of Springer, copyright 2002

References

1. Salikhov KM, Semenov AG, Tsvetkov YD (1976) Electron spin echo and its applications. Nauka, Novosibirsk
2. Salikhov KM, Tsvetkov YD (1977) Electron spin-echo studies of spin-spin interactions in solids. In: Kevan L, Schwartz RN (eds) Time domain electron spin resonance. Wiley Interscience, New York, pp 232–277
3. Milov AD, Tsvetkov YD (2000) Charge effect on relative distance distribution of Fremy's radical ions in frozen glassy solution studied by PELDOR. Appl Magn Reson 18(2):217–226. https://doi.org/10.1007/Bf03162112
4. Syryamina VN, Samoilova RI, Tsvetkov YD, Ischenko AV, De Zotti M, Gobbo M, Toniolo C, Formaggio F, Dzuba SA (2016) Peptides on the surface: spin-label EPR and PELDOR study of adsorption of the antimicrobial peptides trichogin GA IV and ampullosporin a on the silica nanoparticles. Appl Magn Reson 47(3):309–320. https://doi.org/10.1007/s00723-015-0745-5
5. Hilger D, Jung H, Padan E, Wegener C, Vogel KP, Steinhoff HJ, Jeschke G (2005) Assessing oligomerization of membrane proteins by four-pulse DEER: pH-dependent dimerization of NhaA Na^{+}/H^{+} antiporter of *E. coli*. Biophys J 89(2):1328–1338. https://doi.org/10.1529/biophysj.105.062232
6. Milov AD, Samoilova RI, Tsvetkov YD, Peggion C, Formaggio F, Toniolo C (2014) Peptides on the surface. PELDOR data for spin-labeled alamethicin F50/5 analogues on organic sorbent. J Phys Chem B 118 (25):7085–7090. https://doi.org/10.1021/jp503691n
7. Debye P, Hückel E (1923) Zur Theorie der Elecktrolyte. Physikalische Zeitschrift 24(9):185–206
8. Vasserman AN, Kovarskii AL (1986) Spin probes and labels in physical chemistry of polymers. Nauka, Moscow

9. Milov AD, Tsvetkov YD (1997) Double electron-electron resonance in electron spin echo: Conformations of spin-labeled poly-4-vinilpyridine in glassy solutions. Appl Magn Reson 12 (4):495–504
10. Khazanovich TN, Kolbanovsky AD, Kokorin AI, Medvedeva TV, Wasserman AM (1992) E.p.r. spectroscopy of spin-labelled macromolecules as a tool for determining chain conformations in amorphous solid polymers. Polymer 33(24):5208–5214. https://doi.org/10.1016/0032-3861(92)90803-5
11. Wasserman AM, Khazanovich TN, Kasaikin VA (1996) Some EPR spin probe and spin label studies of polymer systems. Appl Magn Reson 10(1):413–429. https://doi.org/10.1007/bf03163122
12. Hinderberger D, Spiess HW, Jeschke G (2004) Dynamics, site binding, and distribution of counterions in polyelectrolyte solutions studied by electron paramagnetic resonance spectroscopy. J Phys Chem B 108(12):3698–3704. https://doi.org/10.1021/jp036043u
13. Pfannebecker V, Klos H, Hubrich M, Volkmer T, Heuer A, Wiesner U, Spiess HW (1996) Determination of end-to-end distances in oligomers by pulsed EPR. J Phys Chem-Us 100 (32):13428–13432. https://doi.org/10.1021/jp960895v
14. Milov AD, Samoilova RI, Tsvetkov YD, Gusev VA, Formaggio F, Crisma M, Toniolo C, Raap J (2002) Spatial distribution of spin-labeled trichogin GA IV in the gram-positive bacterial cell membrane determined from PELDOR data. Appl Magn Reson 23(1):81–95. https://doi.org/10.1007/bf03166186
15. Salton MRJ (1968) Lytic agents cell permeability and monolayer penetrability. J Gen Physiol 52(1p2):227–252
16. Whiteside TL, Desiervo AJ, Salton MRJ (1971) Use of antibody to membrane adenosine triphosphatase in study of bacterial relationships. J Bacteriol 105(3):957–967
17. Thorne KJI, Kodicek E (1962) The metabolism of acetate and mevalonic acid by lactobacilli. IV. Analysis of the fatty acids by gas-liquid chromatography. Biochim Biophys Acta 59 (2):306–312. https://doi.org/10.1016/0006-3002(62)90178-6
18. Salton MRJ (1964) The bacterial cell wall. Elsevier, Amsterdam

Chapter 6
PELDOR in Peptide Research

Abstract The measurement of distances within and between peptides is an important application of PELDOR or DEER spectroscopy in biology. Individual measurements have been made on many proteins and provide distance information that have been vital for understanding the properties of those proteins. However, the most thoroughly studied peptides belong to the class of peptaibols.

Peptides are chains of amino acids linked by amide bonds. They are the major component of proteins and enzymes. The folding of peptides into three-dimensional structures and changes in the conformations of those structures play vital roles in all living organisms. The exact sequence of the different amino acids making up a protein have a profound influence on both structure and biological function. The measurement of distances within and between peptides is an important application of PELDOR in biology. Individual measurements have been made on many proteins and provide distance information that have been vital for understanding the properties of those proteins. However, the class of peptides most-thoroughly studied by PELDOR are the peptaibols.

The site-directed spin labeling of peptides allows PELDOR methods to provide quite detailed structural information on peptides in solution, in aggregates and absorbed on surfaces. The distance distribution spectra can reveal structural heterogeneity and can measure the shifts in population. PELDOR also retains the high specificity and selectivity of EPR for the spin-labeled peptides without the restrictions on optical properties or crystallinity required by other structural techniques.

6.1 Spin-Labeled Peptaibols

The peptaibols are members of a class of largely hydrophobic, membrane-active, linear peptides up to 21 α-amino acids in length and typically having antibiotic activity [1–11]. They often contain the unusual amino acid α-aminoisobutyric acid

© Springer Nature Switzerland AG 2019

Y. D. Tsvetkov et al., *Pulsed Electron–Electron Double Resonance*,
https://doi.org/10.1007/978-3-030-05372-7_6

(Aib), Table 6.1. Unlike the common amino acids, the C^{α} of Aib has four chemical bonds to heavy atoms and none to hydrogen. As a result, the Aib residue is stiffer than other amino acids and strongly foldameric or helicogenic [12, 13], so that peptaibols strongly prefer helical structures in artificial bilayers and in natural membranes where they are bioactive. Peptaibols typically have an N-terminal acyl group, either acetyl or a long chain acyl, and a C-terminal 1,2-amino alcohol which eliminates the terminal positive and negative charges from the backbone. In Nature, most peptaibols exhibit microheterogeneity, i.e., they are a mixture of very similar peptides with numerous conservative changes in their amino acid sequence.

Knowledge about the 3D-structure and properties of peptaibols and their aggregates is necessary for the design of efficient antibiotics based on peptaibols. PELDOR studies have revealed several structural peculiarities, including aggregation of peptaibols into a range of different structures in membrane-mimicking solvents, in lipid model systems, as well as in bacterial cells. Structural details are obtained from the dipole-dipole interactions between spin labels introduced at specific sites in the peptaibol. The TOAC nitroxyl spin label, **4-22**, [14] is typically used for PELDOR studies of peptaibols, Table 6.1. TOAC is covalently and rigidly embedded in the peptide sequence, replacing an Aib and is equally helicogenic. TOAC eliminates the flexible side-chain linker to the nitroxyl group that is inherent in every other amino acid spin label. Consequently, TOAC produces very accurate measurements of the distance between spins, although TOAC's six-membered ring may adopt at least two conformations [15–17].

Formation of pores in biological membranes eventually generates cell leakage and is generally believed to be a basis for the antibiotic activity of membrane-active peptides. Peptaibols are intriguing because many of them can be activated by a membrane voltage [1, 2, 4, 9–11, 18, 19] even though they are electrically neutral. But their ability to cause membrane leakage in the absence of a membrane voltage is even more challenging to explain.

PELDOR provides extremely useful information on the overall 3D-structure and self-association modes of biomolecules in membranes [20]. Other spectroscopic techniques, such as fluorescence, and in particular, fluorescence resonance energy transfer, FRET, can provide complementary information to confirm and further refine the picture provided by PELDOR data.

6.1.1 Secondary Structure

A variety of site-directed mono- and double-labeled peptaibols with various main-chain lengths, Table 6.1, have been studied. The PELDOR experiments discussed here were made with spin-labeled peptaibols prepared by chemical synthesis. A CW EPR approach to correlate peptide conformation with distance between two TOAC labels has been suggested, but seems limited to short distances <1.5 nm [15–17]. PELDOR experiments were performed at X-band on the homemade ICKC or commercial E580 pulsed EPR spectrometers using 3p- or

Table 6.1 Amino acid sequences of peptaibols investigated by PELDOR spectroscopy, bold characters indicate **Aib** residues potentially replaced by TOAC spin label **4-22** in the labeled analogs, with labeled sites indicated by superscripts

Abbreviation	Name sequence[a]
O-Tri	Trichogin GA IV[c] Oct·**Aib**$^{\mathbf{1}}$·Gly·Leu·**Aib**$^{\mathbf{4}}$·Gly·Gly·Leu·**Aib**$^{\mathbf{8}}$·Gly·Ile·Leu·OMe
F-Tri	Trichogin GA IV[b] Fmoc·**Aib**$^{\mathbf{1}}$·Gly·Leu·**Aib**$^{\mathbf{4}}$·Gly·Gly·Leu·**Aib**$^{\mathbf{8}}$·Gly·Ile·Leu·OMe
Tyl	Tylopeptin B[b] Ac·Trp·Ala·**Aib**$^{\mathbf{3}}$·Aib·Ala·Gln·Ala·**Aib**$^{\mathbf{8}}$·Ser·Aib·Ala·Leu·**Aib**$^{\mathbf{13}}$·Gln·Lol
Hep	Heptaibin Ac·Phe·**Aib**$^{\mathbf{2}}$·Aib·Aib·Val·Gly·Leu·Aib·Aib·Hyp·Gln·Aib·Hyp·**Aib**$^{\mathbf{14}}$·Phl
Amp	Ampullosporin A[b] Ac·Trp·Ala·**Aib**$^{\mathbf{3}}$·Aib·Leu·Aib·Gln·Aib·Aib·Aib·Gln·Leu·**Aib**$^{\mathbf{13}}$·Gln·Leu·OMe
Zrv	Zervamicin IIa[b] $\mathbf{T^N}$·Trp·Ile·Gln·Aib·Ile·Thr·Aib·Leu·Aib·Hyp·Gln·Aib·Hyp·Aib·Pro·$\mathbf{T^C}$
Alm	Alamethicin F50/5[b] Ac·**Aib**$^{\mathbf{1}}$·Pro·Aib·Ala·Aib·Ala·Glu(OMe)·**Aib**$^{\mathbf{8}}$·Val·Aib·Gly·Leu·Aib·Pro·Val·**Aib**$^{\mathbf{16}}$· Aib·Glu(OMe)·-Glu(OMe)·-PheOMe·Ac
D-Tri	Trichogin GA dimer[b] Ac·**Aib**$^{\mathbf{1}}$·Gly·Leu·Aib·Gly·Gly·Leu·Aib·Glu·Ile·Leu·Aib·Gly·Leu·Aib·Gly·Gly·Leu·**Aib**$^{\mathbf{19}}$· Gly·Ile·Leu·OMe

[a]Other abbreviations used: Oct, 1-octanoyl; OMe, methoxy; Fmoc, 9-fluorenyimethoxycarbonyl; Ac, acetyl; Lol, leucinol; Hyp, (4R)-hydroxy Pro; Phol, phenylalaninol; Glu(OMe), γ-methyl Glu
[b]Analog

4pPELDOR. Solid glassy samples in organic solvents were measured at cryogenic temperatures, usually 77 K. Additional experimental details are given in the text or in Chap. 2.

In almost all cases, one or two Aib residues [12, 13] were replaced by equally helicogenic TOAC spin labels [14], Table 6.1. The exceptions include two zervamicin analogs labeled with the TEMPO **4-19** derivatives $\mathbf{T^N}$ and $\mathbf{T^C}$ [21, 22], Table 6.1, attached through an amine or carbonyl group at their N- or C-terminus, respectively.

In polar solvents, such as alcohols, peptaibols are monomeric, but usually self-aggregate in less polar solvents. This offers an opportunity to study their 3D-structure and supramolecular aggregation independently. Information about the secondary structure of peptaibols in the monomeric state as well as in the aggregated state can be obtained from PELDOR spectroscopy of doubly spin-labeled peptides. The global quaternary structure of the aggregates can be obtained from mono spin-labeled peptides.

The determination of the peptide conformation by PELDOR uses doubly-labeled peptide dissolved as monomers in a glassy polar solution. The modulation of the PELDOR signal provides a distance distribution spectrum $F(r)$ between the two labels as described in Sects. 1.2 and 1.3. The peptide conformation is determined

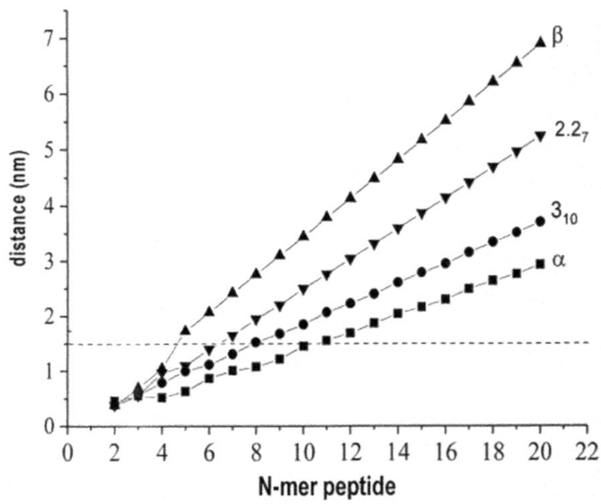

Fig. 6.1 Calculated distances between C^{α} atoms of the terminal residues of *N*-mer peptides for standard α-, 3_{10}-, and 2.2_7-helix and β-sheet conformations: PELDOR measurements of distances are reliable with an accuracy of ±0.03 nm between about 1.5 (*dashed line*) and 8 nm; qualitative information for shorter distances can be obtained by CW EPR spectroscopy. Reproduced from Milov et al. [22] with permission of John Wiley and Sons, copyright 2007

by comparing the measured $F(r)$ with calculated distances between label sites for models of the possible peptide conformations.

The conformational space of peptides is restricted to a range of ϕ and ψ backbone torsion angles, e.g., on the Ramachandran plot, that define α-, 3_{10}-, and 2.2_7-helices, and β-sheet conformations [12, 13, 23–27]. Several helices are possible, each having different patterns of H-bonding between C=O and N–H groups of the peptide backbone. In proteins, 3_{10}-helices have been proposed as kinetic intermediates in the folding and unfolding of long α-helices [28, 29]. In heptapeptides, these two helix conformations occur with roughly equally probability [12, 13, 23, 27]. Many of the very short, Aib-rich peptides prefer the 3_{10}-helix.

The H-bonding patterns produce different lengths for a given peptide, with length increasing in the order α- < 3_{10}- < 2.2_7-helix < β-sheet. This link between conformation and length is the crucial relation that allows peptide conformation to be determined by PELDOR. Distances between terminal residues are calculated for peptides of increasing length in different conformations in Fig. 6.1 [22].

Trichogin GA IV

CW EPR spectra of doubly spin-labeled trichogin GA IV **O-Tri** and **F-Tri** have a weak half-field transition near $g = 4.0$ in frozen glassy solutions [17]. This formally 'forbidden' transition becomes partially allowed by the dipolar interactions between spin labels at fairly short distances. Based on the intensity of this line, the conformation of this short lipopeptide with eleven amino acid residues was assigned as a mixed α/3_{10}-helix, as found by X-ray diffraction in crystals of trichogin GA IV and **O-Tri^{4,8}**, Table 6.1 [7, 8]. However, PELDOR data reveals that the conformation of **O-Tri^{1,8}** depends on the nature of its solvent, Table 6.2 [30–32]. In frozen methanol (MeOH) or in chloroform/dimethylsulfoxide ($CHCl_3$/DMSO) (7:3) the experimental interspin distance, $r = 1.97$ nm, corresponds to a mixed $2.2_7/3_{10}$-helix, while in 2,2,2-trifluoroethanol (TFE), the distance is shorter, $r = 1.53$ nm, corresponding to a pure 3_{10}-helix [31]. Although the ribbon-like 2.2_7-helix is seldom seen [27] and it is difficult to discriminate between 2.2_7- and 3_{10}-helices shorter than $r = 1.70$ nm, Fig. 6.1, both results are supported by CW EPR experiments [15].

PELDOR can determine the conformation of peptides even in the aggregated state. The double spin-labeled peptide must be diluted with its unlabeled counterpart to minimize intermolecular dipole-dipole interactions. This strategy showed that **O-Tri^{1,8}** has a 3_{10}-helical conformation in the aggregated state [32]. Short peptaibols, such as trichogin, have conformational flexibility, and their structure depends on temperature and on the nature and organization of their surroundings, e.g., glassy state, crystalline state, etc.

Tylopeptin B, Heptaibin, and Ampullosporin A

The distance distribution spectrum $F(r)$ of **Tyl^{3,13}**, Table 6.1, a 14-amino acid analog of tylopeptin B, has a single Gaussian-shaped peak, $r_{max} = 1.76\,\text{nm}$, Fig. 6.2a and Table 6.2 [35, 39]. The main part of the spectrum for **Hep^{2,14}**, a 14-amino acid residue analog of heptaibin also has a Gaussian shape, but it is broader and shifted to longer intramolecular distances, $r_{max} = 2.3\,\text{nm}$, Fig. 6.2a and Table 6.2 [35, 39]. These results indicate that **Hep^{2,14}** has a 3_{10}-helical conformation, while **Tyl^{3,13}** seems to adopt a largely α-helical conformation. These two peptaibols have the same number of amino acids, although they adopt different conformations. One significant difference between these two peptaibols is that only the heptaibin analog contains the dipeptide sequence Aib-Hyp which particularly favors the more elongated 3_{10}-helix.

The spin-labeled **Amp^{3,13}** analog of ampullosporin A is another 14-amino acid residue peptaibol, which was studied both by CW EPR and PELDOR [40] in glassy MeOH at 77 K. From PELDOR experiments, the distance spectrum exhibits a main, Gaussian line with a peak at $r_{max} = 1.73\,\text{nm}$, Table 6.2. Similar results were found from CW EPR analysis and by using convolution/deconvolution methods [41, 42]. The interspin distance of 1.73 nm suggests that **Amp^{3,13}** adopts a mainly α-helical structure, in agreement with CD and fluorescence data in the same paper [40].

Zervamicin II A

Zervamicin II A is a peptaibol with an additional amino acid residue, for a total of 15 residues. X-Ray diffraction and NMR analysis indicated that it folds in a mixed conformation with an α-helix at the N-terminus and a 3_{10}-helix at the C-terminus [28, 43, 44]. The **T^N-Zrv-T^C** analog with TEMPO spin labels attached at each end, was studied by PELDOR in frozen solutions of MeOH, MeOH/toluene, and MeOH/$CHCl_3$ [21, 22]. In MeOH, the main maximum of the distribution function lies at a distance of ~3.2 nm, Table 6.2. This distance appears to depend only slightly on

Table 6.2 Dependence of conformation[a] on solvent polarity and oligomeric number for double spin-labeled peptaibol analogs of different length [33]

Peptaibol analog	Solvent	Oligomeric number (N)	Distance, *r* (nm)	Δ (nm)	Helix type	References
O-Tri[1,8]	$CHCl_3$/toluene 7:3	4	1.57		3_{10}	[32]
O-Tri[1,8]	$CHCl_3$/DMSO, MeOH	1	1.97		2.2_7/3_{10}^b	[31]
O-Tri[1,8]	TFE	1	1.53		3_{10}	[31]
Tyl[3,13]	MeOH/ethanol 95:5	2	1.76	0.07	α	[34, 35]
Hep[2,14]	MeOH/ethanol 95:5	2	2.30	0.07	3_{10}	[34, 35]
Amp[3,13]	MeOH	1	1.73	0.05	α	[35]
T[N]-Zrv-T[C]	MeOH	2	3.20	0.3[c]	α/3_{10}^b	[21, 22]
Alm[1,16]	MeOH	8	2.1	0.3[c]	α	[36]
Alm[1,16]	MeOH/toluene 1:4	8	2.1	0.3[c]	α	[36]
D-Tri[1,19]	MeOH	1	2.8/3.2	2	α, 3_{10}^b	[37, 38]
D-Tri[1,19]	MeOH/toluene, $CHCl_3$/toluene	2	2.8	0.3	α	[37, 38]

[a]Determined for the segment $Aib^i - Aib^j$ from the $TOAC^i$ to $TOAC^j$ distances at 77 K
[b]The notations "$2.2_7/3_{10}$" and "α/3_{10}" indicate mixed conformations within *each* molecule, while "α, 3_{10}" is used for a mixture of molecules with *different* conformations
[c]Asymmetric distance distributions due to the slightly increased flexibility of the TEMPO labels $\mathbf{T^C}$ and $\mathbf{T^N}$

the solvent composition, but the distribution function narrows after addition of either $CHCl_3$ or toluene to the MeOH. This effect was rationalized in terms of decreased mobility at both termini. Molecular dynamics simulations showed that the dominant PELDOR distance agrees well with the mixed α/3_{10}-helix previously determined by NMR. However, when toluene was added to the MeOH solution to further increase the hydrophobicity of the solvent, a minor fraction (7–8%) appears with a distance of 4.2 nm. Most likely, this conformation corresponds to the more elongated 2.2_7-helix for this membrane-active peptide.

Alamethicin F50/5

The double spin-labeled **Alm**[1,16], Table 6.1, is an analog of the long, 19-amino acid peptaibol alamethicin F50/5 [45]. It was investigated in frozen solutions of MeOH and MeOH/toluene (1:4), Table 6.2 [36]. The $r_{max} \sim 2.1$ nm distance found in both solvents was in excellent agreement with average distances calculated from a set of five models based on the almost-completely α-helical crystal structures [5, 46]. The distance between labels corresponds to an α-helix, at least for the segment between the labels at residues 1 and 16. Compared to trichogen which has different conformations in different solvents, alamethicin has a remarkably stable conformation. In particular, its conformation remains constant as polarity changes from MeOH to MeOH/toluene.

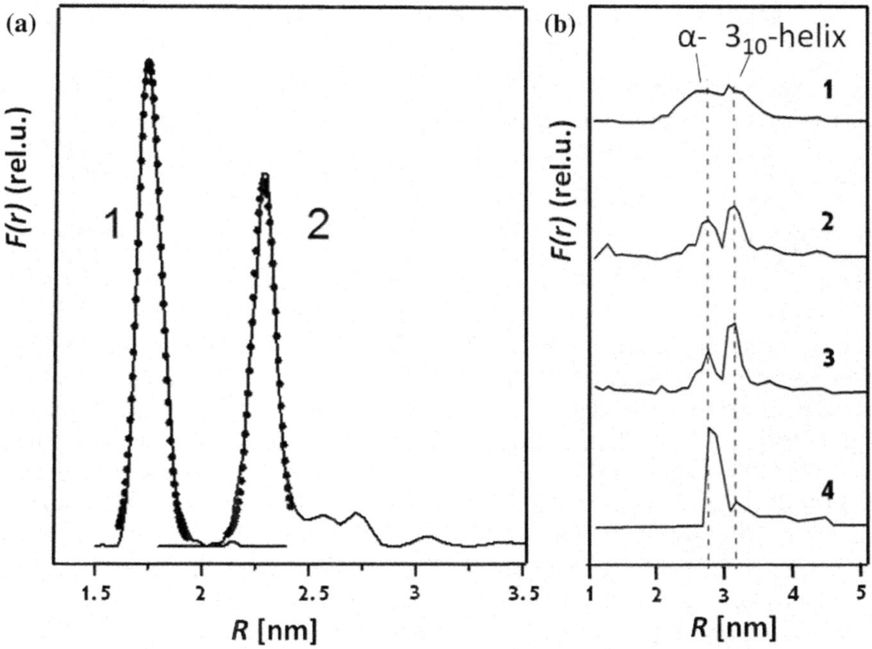

Fig. 6.2 Conformation and environment dependencies: **a** intramolecular distance distributions between spins in the doubly spin-labeled peptaibol analogs *1* **Tyl**^{3,13} and *2* **Hep**^{2,14} agree with a 3_{10}- and an α-helix conformation, respectively, see Fig. 6.1; **b** changes in the distance distribution function $F(r)$ of **D-Tri**^{1,19} with solvent polarity: *1* MeOH; *2* MeOH/toluene 10%; *3* MeOH/toluene 20% MeOH; and *4* chloroform/ toluene. Reproduced from Milov et al. [39] with permission of Springer, copyright 2013 and from Milov et al. [33]

Covalent Dimer of Trichogin GA IV

The longest peptaibol examined by PELDOR is a 22-amino acid, covalently-linked dimer of Trichogin GA IV, designed and synthesized explicitly for PELDOR to test models of aggregation. The distance distribution functions obtained for the doubly spin-labeled **D-Tri**^{1,19}, Table 6.1, in MeOH/toluene mixtures, Fig. 6.2b, reveals the effect of solvent on self-aggregation and the concomitant alteration of secondary structure [37, 38]. In pure MeOH, the PELDOR time trace $V(T)$ has weak oscillations and $F(r)$ has a poorly-resolved doublet nearly 2 nm wide, Table 6.2 Addition of toluene increases the oscillation amplitude and $F(r)$ exhibits two distinct maxima at $r_{max} = 2.8$ and 3.2 nm corresponding to α-helix and 3_{10}-helix conformations, respectively, Fig. 6.2b. In pure toluene, $V(T)$ is strongly modulated and the main $F(r)$ peak corresponds to α-helical, aggregated molecules.

6.1.2 Quaternary Structures in Frozen Solutions

Spin-labeled peptaibols aggregate in solvent mixtures with low polarity. Their quaternary structures, or arrangement of peptides within an aggregate, are obtained by PELDOR spectroscopy from dipolar interactions between labels in different peptides.

Trichogin GA IV

The transition from polar to nonpolar solvents changes the conformation of trichogins and causes aggregation [47]. In nonpolar media, both **F-Tri[4]** and **O-Tri[1]** mono spin-labeled analogs show a fast drop in their $V(T)$ at small T, followed by a slow decay modulated by the dipolar interaction, Fig. 6.3, curves *1* and *2*. Addition of the polar EtOH to the same samples changes the decays to exponential functions, Fig. 6.3, curves *3* and *4*, indicating a disruption of the aggregates into monomers. For both **F-Tri[4]** and **O-Tri[1]**, $N = 4 \pm 0.3$ spin labels were found in each aggregate and the distances, within an aggregate, between spin labels on different monomers were $r = 2.35$ nm and 2.60 nm.

Similar aggregation phenomena were found in other frozen glassy solutions: $CHCl_3$/toluene, $CHCl_3$/decalin, CCl_4/toluene, and dichloroethane/toluene. The exact quaternary structure that forms depends somewhat on the solvent properties, with r_{max} ranging between 2.3 and 3.3 nm, and N between 3.1 and 4.3 [48]. Doubly spin-labeled trichogin analogs gave consistent results [32]. A model of the aggregate that is consistent with the experimental results contains four peptides arranged as a pair of dimers, Fig. 6.4 [47]. The peptaibols in each dimer are antiparallel. In nonpolar solvents, each peptaibol is a 3_{10}-helix. The tetramer is formed by two dimers. The distances between labels on different peptides are about 2.3 and 2.6 nm, in aggregates of **O-Tri[4]** and **O-Tri[1]**, respectively. These findings agree well with a helix bundle where the helices are antiparallel. This model of the

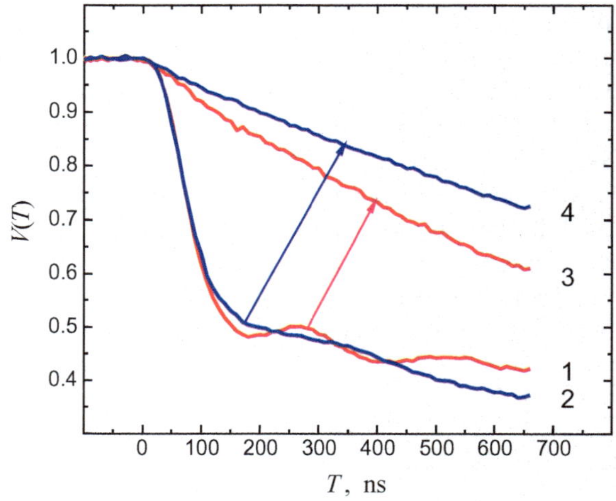

Fig. 6.3 $V(T)$ for mono spin-labeled **F-Tri[4]** and **O-Tri[1]** in $CHCl_3$/toluene, curves *1* and *2*, respectively, and after addition of EtOH, curves *3* and *4*, respectively; measurements were made at 77 K. Reproduced from Milov et al. [47] with permission of American Chemical Society, copyright 2000

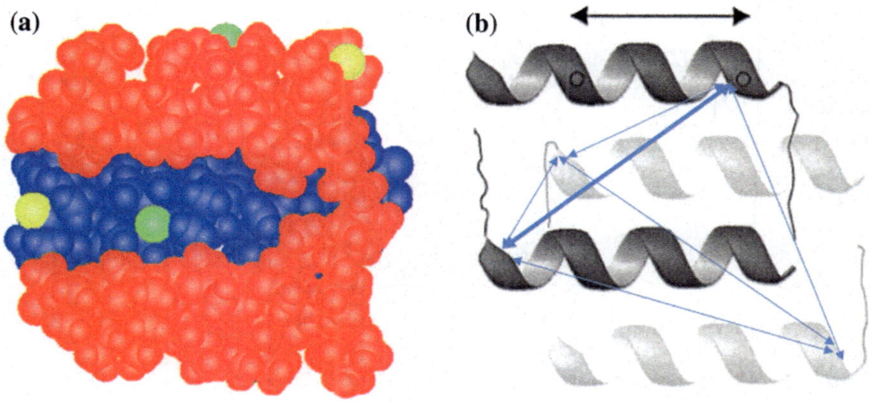

Fig. 6.4 **a** A space-filling molecular model of the quaternary structure of the aggregate of the double spin-labeled **O-Tri**[1,4] analog in nonpolar media; **b** A ribbon drawing of the supramolecular assembly of one double spin-labeled **O-Tri**[1,4] and three unlabeled **O-Tri** peptide chains; for clarity, the distances are only shown for the TOAC[1] positions. Reproduced from Milov et al. [47] with permission of American Chemical Society, copyright 2000

tetramer resembles a vesicle: polar sidechains point to the interior of the tetramer and the nonpolar sidechains to the exterior.

This tetramer model, based on PELDOR experiments, was corroborated by CW EPR data [49, 50]. Two different sets of EPR signals were observed at room temperature in the $CHCl_3$/toluene solvent mixture, which was attributed to an equilibrium between monomers and aggregates. A two-step aggregation mechanism was proposed: dimerization of the peptide molecules, followed by aggregation of the dimers into tetramers. The estimated equilibrium dissociation constant for the first step is ~ 17 μM and for the second step ~ 1 μM. The EPR line widths depend on the rotational mobility of the aggregates and the monomers, which depend on their sizes. Analysis of the line widths gave an estimate of 4 peptide molecules per aggregate,

Tylopeptin B and Heptaibin

The aggregation of tylopeptin B was investigated using tylopeptin B analogs with spin labels at position 3, 8 or 13, Table 6.3. In weakly polar MeOH/toluene solutions, two or three peptide molecules were estimated to make up the aggregates, based on the concentration dependence of the modulation depth [35, 39]. Dimer formation was suggested to be the prevailing mode of self-association with molecules arranged head-to-head. The width of the distance spectrum with the label at the C-terminus in **Tyl**[13] seems much broader than for **Tyl**[3] with the label at the N-terminus, suggesting that the C-terminus is more mobile than the N-terminus.

Both head-to-head and head-to-tail dimerization was suggested in heptaibin analogs labeled at position 2 or 14 [35, 39], Table 6.3. The distance distribution spectra contained broad lines, suggesting aggregates containing 3 or 4 peptide molecules, in addition to the dimers.

Table 6.3 Quaternary structures of mono spin-labeled peptaibols in solvent mixtures of low polarity [33]

Peptaibol analog	Label positions	Solvent	N^a	Packing mode	References
O-Tri	1, or 4, or 8	$CHCl_3$/toluene	4	Antiparallel	[32, 47]
Tyl	3 or 13	MeOH/toluene	2	Head-to-head[b]	[35, 39]
Hep	2 or 14	MeOH/toluene	2	Head-to-tail and head-to-head[b]	[35, 39]
Zrv-T^C	C-term	MeOH/toluene	2	Antiparallel	[21, 22]
Alm	1, or 8, or 16	$CHCl_3$/toluene	8	Parallel dimer of a linear tetramer	[51, 52]
D-Tri	1, 19	$CHCl_3$/toluene	2	Antiparallel	[37, 38]

[a]Oligomeric number
[b]Probably a mixture of aggregates of different types

Zervamicin II A

Aggregation of mono spin-labeled zervamicin **Zrv-T^C** molecules was studied in toluene/MeOH mixtures containing 8.5–16% MeOH [21, 22]. The PELDOR $V(T)$ shows a fast decay to a constant limiting value of V_p without any oscillations. This type of trace is characteristic of clusters of electron spins. One can estimate the effective distance r_{eff} between spins in the group from the decay, Sect. 1.2.6. Values of r_{eff} lie between 2.5 and 3.5 nm, depending on solvent. NMR experiments indicate that helical zervamicin molecules in solution have a length of 2.56 nm [6]. Thus, the distance between spin labels at the C-termini of **Zrv-T^C** in the aggregate is consistent with antiparallel arrangements of peptaibols. Moreover, the mean value of $\bar{N}$ calculated from Eq. 1.27 depends on the solvent mixture with $\bar{N} = 2$ for at least 44–67% of **Zrv-T^C** molecules. The sparse 3D-structural information for zervamicin prevents the construction of a detailed molecular model for the aggregate of this peptaibol.

Alamethicin F50/5

Alm analogs spin-labeled at positions 1, 8, or 16 were investigated by CW EPR and PELDOR techniques [51, 52]. The $F(r)$ functions, Fig. 6.5a, from each of the three peptides in $CHCl_3$/toluene (1:1) at concentrations of 10^{-3}–10^{-4} M were remarkably similar. The $F(r)$ functions have a clear, sharp maximum at 3.18 nm for **Alm^1**, 3.24 for **Alm^8**, and 3.06 for **Alm^{16}**, and a much broader maximum near 7 nm. Estimation from the area under regions in *F(r)*, Fig. 6.5a, shows that nearly 30% of the spin labels have the short distances and the remaining 70% have the larger distances. The aggregates seems to contain $N = 6–8$ mol.

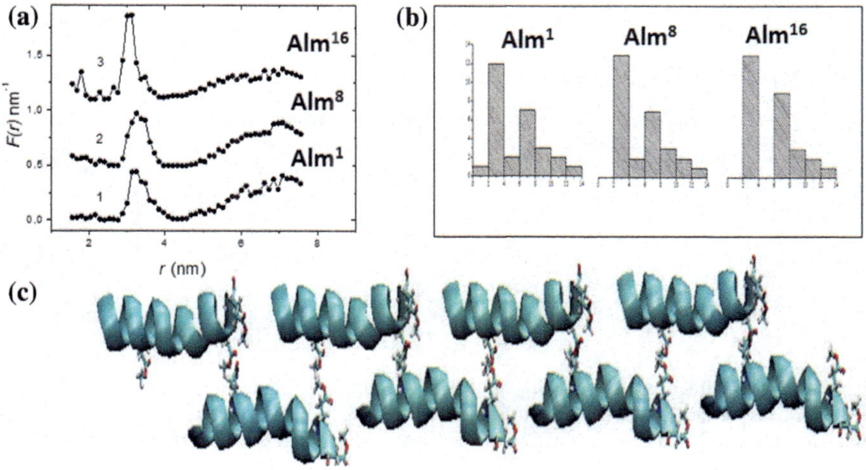

Fig. 6.5 **a** Distance distributions $F(r)$ for mono spin-labeled **Alm¹**, **Alm⁸**, and **Alm¹⁶** aggregates in $CHCl_3$/toluene (1:1) at 77 K, curves *2* and *3* are shifted upwards by 0.5 and 1.1, respectively, to aid comparison; **b** histograms of the distances between labels from MD simulations of an ocatmer model for **Alm¹**, **Alm⁸**, and **Alm¹⁶**; the total number of distances determined for each aggregate is 28; **c** a molecular model of an aggregate of **Alm** as an octamer formed as two parallel chains, each composed of four head-to-tail monomers. Reproduced from Milov et al. [51] with permission of John Wiley and Sons, copyright 2007

The CW EPR line shapes of these peptides were studied in chloroform/toluene mixtures at room temperature to obtain a more precise value for N. From the rotational correlation times, N was estimated to be 6.8 ± 2.5 [53]. The combined PELDOR and CW EPR results indicates that $6 \leq N \leq 9$. A model for the **Alm** aggregate was proposed as two linearly aligned tetramers, Fig. 6.5c.

Detailed MD calculations were carried out to test the proposed model. The minimum peptide-peptide interaction energy occurs for molecular conformations in which the Glu γ-ester groups at positions 7, 18, and 19 in both peptides interact as dipoles with antiparallel orientations, thus forming seven polar clusters along the alamethicin aggregate, Fig. 6.5c. This model places the nitroxyl groups on adjacent monomers about 3.3 nm apart. Histograms of the distances between labels, Fig. 6.5b, for this model are in qualitative agreement with the experimental $F(r)$ for the mono spin-labeled **Alm¹**, **Alm⁸**, and **Alm¹⁶**.

Dimer of Trichogin GA IV

In this peptide, the sequence of trichogin repeats twice to form a head to tail, covalently-linked dimer, rather than an antiparallel dimer seen in many of the peptaibol aggregates that form in solution. This makes it the longest peptaibol ($n =$ 22) studied by PELDOR [37, 38]. The aggregate of the mono spin-labeled **D-Tri¹** has $N \sim 2$, based on V_p, and a distance of 3.4 nm between the spin labels of different molecules. This result implies an antiparallel arrangement of molecules in the aggregate. The estimated total dipole moment of the **D-Tri¹** α-helix is 77.0

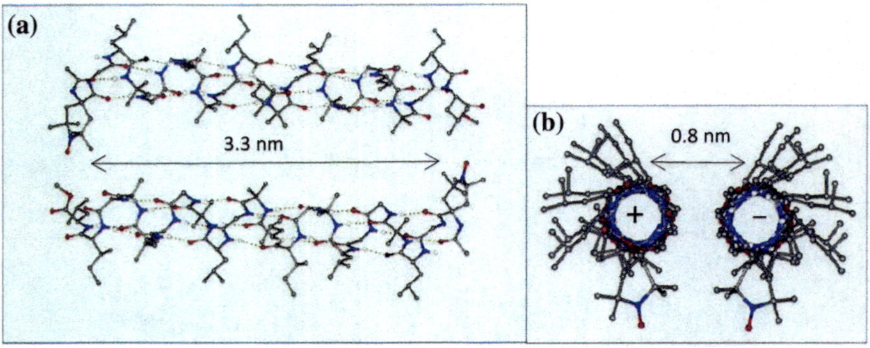

Fig. 6.6 **a** 3D-Structural model of the **D-Tri**[1] aggregate: molecules with an antiparallel orientation and a label-label distance in the dimer of 3.3 nm and **b** a view along the axes of the α-helices; the centers of the two helices are separated by about 0.8 nm. Reproduced from Milov et al. [37] with permission of American Chemical Society, copyright 2003

Debye, aligned along the helix. This large electric dipole strongly favors the antiparallel arrangement, while the distance between α-helices is determined only by steric factors. A 3D-structural model for the **D-Tri**[1] peptide aggregate was based on these results, Fig. 6.6.

6.1.3 Secondary Structures in Membrane Systems

Trichogin GA IV

Trichogin GA IV has a remarkable ability to modify cell membranes even though its length is much less than the thickness of a membrane [1, 2, 4]. Detailed information about its aggregation state in the membrane is needed to understand how this short peptide produces such a dramatic effect on membrane permeability. The experimental distance distribution function of **F-Tri**[1] aggregates in a model membrane composed of DPPC (1,2-dipalmitoyl-*sn*-glycero-3-phosphocholine) lipid bilayers exhibits a maximum near 2.5 nm and a half width of 1.4 nm, Fig. 6.7. This is a broad distance distribution, suggesting that several different types of aggregates form. At 5 mol% peptide relative to the DPPC in the bilayers, aggregates are mostly pairs of trichogin molecules with $N = 2.1 \pm 0.1$. In other solvents, N for the spin-labeled trichogin was 1, i.e., a monomer, in the polar solvent EtOH or 4.3, a tetramer, in nonpolar dichloroethane/toluene.

Experiments with fluorescent trichogin analogs at much lower concentrations in large unilamellar vesicles, or LUVs, composed of egg phosphatidyl choline (ePC) and cholesterol gave a lower limit of $N = 2.3$ mol per aggregate [54], while the rate-limiting step for ion conduction in ePC LUVs with cholesterol involves 3–4 unlabeled trichogin molecules at a similar concentration [55]. For peptide

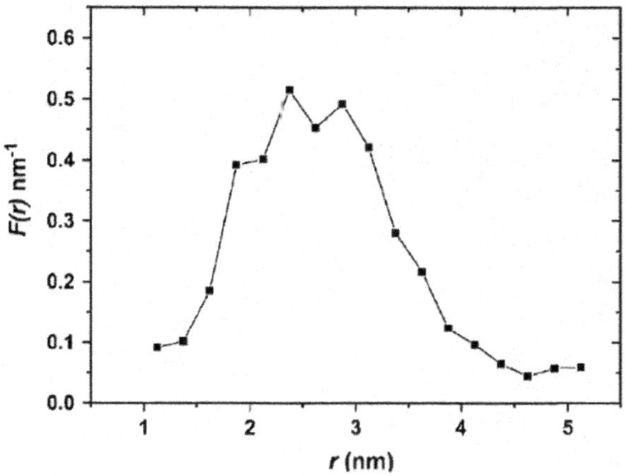

Fig. 6.7 Distance distribution spectra $F(r)$ for spin labels of **F-Tri**[1] aggregates in DPPC lipid bilayers. Reproduced from Milov et al. [56] with permission of Springer, copyright 2005

concentrations of 0.5–2.2 mol% in ePC vesicles, PELDOR studies showed that **O-Tri**[4] is distributed homogeneously in the membranes [56, 57], but the distribution became inhomogeneous and aggregates formed upon addition of cholesterol. The cholesterol seems to stabilize aggregates in the model membranes.

Crystallographic data for trichogin GA IV shows a mixed 3_{10}-/α-helix structure for the peptide backbone [7, 8]. Since the α- and 3_{10}-helices have different lengths, PELDOR can determine the conformation in the membrane from the distance between labels on the same molecule [37, 38, 56–58]. The distance distribution function $F(r)$ was measured for the double spin-labeled **F-Tri**[1,8] in ePC membranes, Fig. 6.8a, [56]. Two main maxima occur at distances of 1.25 and 1.77 nm, corresponding to the α- and 3_{10}-helical conformations, respectively. Apparently, each trichogin molecule is in one of two conformations, with the α-helix being more common. The distance distribution function for **O-Tri**[1,8] in DPPC membranes [58] also corresponds to a mixture of α- and 3_{10}-helix populations, but with the 3_{10}-helix favored, Fig. 6.8b.

In both ePC and DPPC membranes, the trichogins were monomers in the PC head-group regions of the membranes [58]. Thus, the different ratios of helix populations in the two membranes cannot be caused by a different molecular environment, e.g., polar head group versus hydrophobic core, or by a different aggregation state. The trichogin analogs used with the two membrane models are capped at the N-terminus by different groups, i.e., the Fmoc group in **F-Tri**[1,8] and the Oct group in **O-Tri**[1,8]. The Fmoc is a fused, rigid aromatic moiety, while the Oct group has a long, very flexible hydrocarbon chain. Capping a peptide on the N-terminus is well-known to stabilize the helical conformation [27], but the physical basis for the dramatically different helix populations in Fig. 6.8 remains unclear and deserves further investigation.

The spatial distribution of the mono-labeled trichogin GA IV **O-Tri**[4] in spherical cell membranes of the Gram-positive bacterium *Micrococcus luteus* are discussed in Sect. 5.4.

Fig. 6.8 $F(r)$ obtained for **a** **F-Tri**^{1,8} in ePC; and **b** **O-Tri**^{1,8} in DPPC membranes. Reproduced from Milov et al. [58] with permissions of Royal Society of Chemistry, copyright 2004

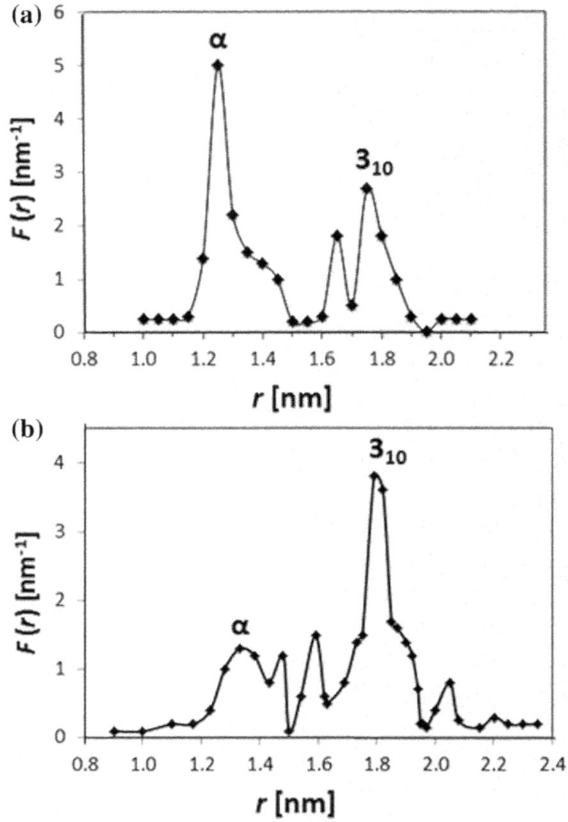

Alamethicin F50/5

Self-assembled aggregates of **Alm** were studied in vesicle membranes at peptide to lipid molar ratios in the range 1:70–1:200. The intermolecular distance distribution was obtained from mono spin-labeled peptides at 77 K, and the aggregation number was estimated as $N \approx 4$ [59–62]. The distance distributions have similar maxima near 2.3 nm. **Alm**^{16} had no other peaks, while **Alm**^{1} and **Alm**^{8} had additional maxima at 3.2 and ~5.2 nm, Fig. 6.9.

Two important features are that these distributions are independent of the spin label position, and they have maxima of ~2.3 nm. These agree well with a simple model of parallel, face-to-face helices with polar groups oriented to the inside and the spin labels to the outside of the aggregate. But these results are inconsistent with an antiparallel orientation of helices which would place the TOAC labels of **Alm**^{16} 3.4 nm apart, well beyond the experimental $F(r)$ peak, Fig. 6.9 curve *3*.

A molecular modeling study was carried out to gain greater insight into the helix-helix interactions [62]. The modeling used the intermolecular PELDOR $F(r)$ between TOAC labels as distance restraints. Each peptide molecule was modeled

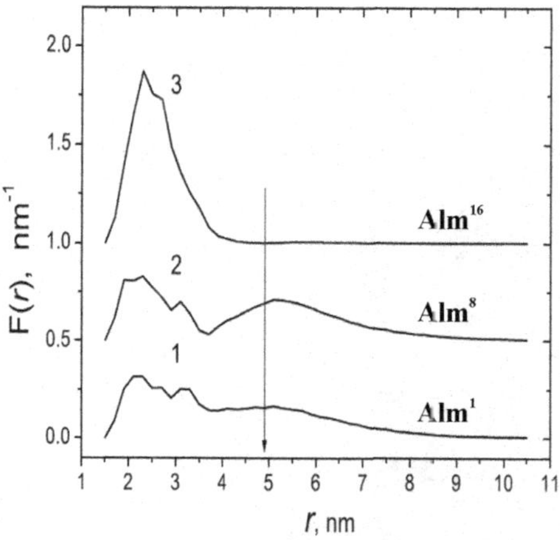

Fig. 6.9 Distance distribution $F(r)$ between TOAC spin-labels for **Alm** analogs in membranes of frozen ePC vesicles; for clarity, curves *2* and *3* are shifted upward by 0.5 and 1.0, respectively, the shape of $F(r)$ at distances longer than that indicated by the arrow is not real and can be used only for estimation of the area under the broad line. Reproduced from Milov et al. [62] with permissions of Elsevier, copyright 2009

with TOAC labels at positions 1, 8, and 16, so that only a single structure had to be optimized but distances could be calculated for all three mono-labeled analogs. The gas-phase, energy-minimized peptide complex has four parallel, but slightly tilted, α-helices, Fig. 6.10. The α-helical structure is broken near the center of each peptide by Pro^{14}. A similar break was also found in a ^{1}H-NMR study of micelle-associated **Alm** [63].

The intramolecular distances between the TOAC residues of **Alm^{1,16}** were not constrained in the model, and confirmed the α-helical conformation in the aggregate [62]. An earlier PELDOR study of doubly spin-labeled **Alm^{1,16}** aggregates in ePC membranes found an average distance of 2.1 nm with a deviation of 0.5 nm between intramolecular labels at peptide to lipid ratios of 1:50 or 1:200 [36]. This experimental distance agrees with the model within experimental error. Thus, the membrane-bound peptides are α-helical despite the disruption by Pro^{14}. Close examination shows that the intermolecular distance of 1.8 nm between the nitroxyl oxygen atoms of the **Alm^{1}** residues matches the first peak of the PELDOR experiments, but not the maximum at 2.3 nm. It should not be a surprise that a gas-phase structural model fails to reproduce every detail of a membrane-embedded protein complex.

This model structure can be accommodated in the hydrophobic region of the membrane by orienting the N-terminal helices of the blue and red colored chains with tilt angles of 15°, the cyan chain by 13°, and the yellow by 8° from the membrane normal, Fig. 6.10. Similar tilts were found for **Alm** by EPR and solid-state NMR analyses [64, 65]. The blue, red, and yellow transmembrane helices extend ~3.2 nm, This is a good match to the ~3.5 nm hydrophobic

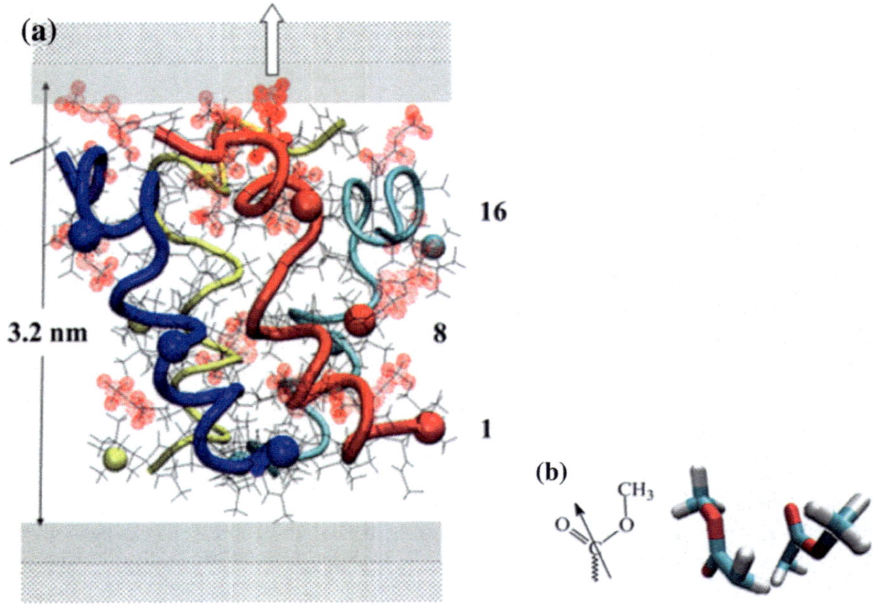

Fig. 6.10 **a** Energy-minimized model of a triply-labeled **Alm** tetramer, based on the short 2.3 nm PELDOR distance between nitroxyl oxygen atoms (indicated by balls) of singly-labeled **Alm**$^{\mathbf{1}}$, **Alm**$^{\mathbf{8}}$, and **Alm**$^{\mathbf{16}}$; numbers on the right-hand side indicate the approximate membrane depth of the corresponding labeled position; each helix is interrupted at Pro^{14}; **b** relative positions of ester groups at the end of Glu^{18}sidechains; for clarity, only one pair is shown; strong electric dipole-dipole interactions between ester groups stabilize the helix bundle and seem to anchor the aggregate to the polar region of the membrane. Reproduced from Milov et al. [62] with permissions of Elsevier, copyright 2009

thickness of the ePC membrane, Fig. 6.10a, in view of the uncertainty in the thickness of the membrane [66, 67] and the possibility of peptide-induced membrane thinning effects [68]. The C-terminus of the cyan chain in this model is immersed more deeply in the hydrophobic region of the membrane.

The polar Glu(OMe) [7, 8, 69] residues of **Alm** seem to have an important role in the closed, or non-conducting, state of the **Alm** channel. Their approximate positions are indicated by dotted circles in Fig. 6.10a, taking into account the flexibility of the long γ-ester side chains. The Glu^{18}(OMe) γ-esters of the red and cyan molecules are only 0.36 nm apart, Fig. 6.10b. Electric dipole-dipole interactions between these γ-ester groups were key in stabilizing aggregates of **Alm** in hydrophobic solvents. It seems that interactions of the C-termini play a similar role in aggregate formation in the ePC membrane. In addition, four of the eight Glu (OMe) groups at the C-terminal are in or near the polar region of the membrane, Fig. 6.10a. This suggests a role for the γ-ester groups of the Glu(OMe) [8] residues in anchoring the aggregate to the polar region of the membrane via dipole-dipole interactions with ester groups of the lipid.

6.1.4 Peptaibols on Surfaces

Peptides absorbed on inorganic or organic surfaces are important in many interdisciplinary areas, such as nanotechnology, biotechnology, medicine, food production, tissue engineering and geochemistry. The properties of absorbed peptides are determined by the primary and secondary structures of the peptides, the nature of the surface, and the type of interaction at the peptide-surface interface.

NMR studies of such systems has achieved some remarkable successes [70, 71]. Solid-state NMR studies of dipolar interactions between magnetic nuclei in a peptide can establish the conformation and orientation of the peptide on different sorbents. Similarly, EPR and PELDOR studies of dipole-dipole interactions in spin-labeled peptides provides information on peptide conformations and aggregation motifs on various types of surfaces. One such PELDOR study of alamethicin absorbed on organic and inorganic sorbents illustrates this.

On HLB

We saw earlier in this chapter that the secondary structure of alamethicin F50/5, or **Alm**, is rather sensitive to the molecular properties of its environment and this holds true for **Alm** on surfaces. Mono and doubly spin-labeled analogs were used as absorbates: $\mathbf{Alm^{1}}$, $\mathbf{Alm^{16}}$, $\mathbf{Alm^{1,16}}$, Table 6.1 [72]. The polymeric, reverse-phase organic sorbent Oasis HLB was the surface. HLB has both hydrophilic and lipophilic properties, and is widely used as the solid phase for liquid chromatography. The peptides were absorbed onto HLB from methylene chloride solution and dried under a flow of dry nitrogen. The amount absorbed was measured at 77 K by CW-EPR.

Intra- and intermolecular contributions to the PELDOR $V(T)$ were separated, taking into account the actual fractal dimensionality of the system. The $V_{\mathrm{INTER}}(T)$ showed a fractal dimensionality d of 2.6 and not the 2.0 expected for an ideal planar surface, Sect. 1.2.4. This indicates that the surface of HLB itself has some fractal character. The $V(T)$ indicate aggregation of **Alm** on HLB. The background–corrected $V(T)$ were analyzed using Tikhonov regularization, Fig. 6.11. Solutions of $\mathbf{Alm^{1,16}}$ in ethanol were used as a convenient reference.

The distance distributions for $\mathbf{Alm^{1,16}}$ in ethanol and on HLB, Fig. 6.11, curves *1* and *2*, respectively, are asymmetric. In ethanol, $r_{max} = 2.00 \pm 0.02$ nm and the width at half-height is $\Delta = 0.25 \pm 0.03$ nm. The distance distribution for the absorbed $\mathbf{Alm^{1,16}}$ is much broader, $\Delta = 0.75 \pm 0.1$ nm, although the peak position is virtually identical, $r_{max} = 2.00 \pm 0.03$ nm [72]. The 2.0 nm distance corresponds to an α-helical conformation. The tail of the peak in ethanol extending to longer distances is probably due to a minor fraction of peptides having elongated $\alpha/3_{10}$-helix conformations with distances in the range $2.0 < r < 2.7$ nm, Fig. 6.11. The fraction of these conformers increases substantially when the peptide is absorbed on HLB. The broadening of the main line with $r_{max} = 2.00$ nm and an increase in the elongated conformers are likely due to the greater diversity in peptide surroundings on HLB as compared to the frozen ethanol.

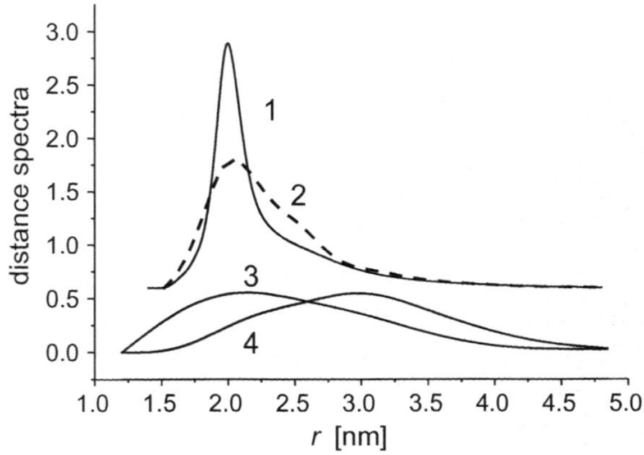

Fig. 6.11 Intramolecular $F(r)$ between TOAC spin labels in **Alm**^{1,16} and intermolecular $F(r)$ between TOAC labels in **Alm** aggregates: Curves *1* and *2* are from **Alm**^{1,16} in ethanol and on HLB, respectively, Curves *3* and *4* are from aggregates of mono-labeled **Alm**^{16} and **Alm**^{1} on HLB, respectively, Curves *1* and *2* are shifted upward by 0.6. Reproduced from Milov et al. [72] with permissions of American Chemical Society, copyright 2014

The $F(r)$ maximum between labels in aggregates on HLB depends on the position of the label. The $r_{max} = 2.1 \pm 0.1$ nm for **Alm**^{16} on HLB shifts to $r_{max} = 3.0 \pm 0.1$ nm for **Alm**^{1}, Fig. 6.11 curves *3* and *4*, respectively. This shift suggests that **Alm** forms aggregates with the C-terminal regions near each other. However, the distance distribution between labels at either position is remarkably broad, $\Delta = 1.6 \pm 0.1$ nm.

On silica nanospheres

Doubly spin-labeled peptaibols have been studied absorbed onto monodisperse silica (SiO_2) nanospheres about 20 nm in diameter. One was the short trichogin, **Tri**^{1,8}, and the other a medium-length ampullosporin, **Amp**^{3,13}, Table 6.1 [73].

The EPR spectra of **Tri**^{1,8} broadened upon absorption, showing that trichogin forms closely-packed peptide clusters. Washing the surface with excess solvent, dispersed the clusters into non-aggregated, surface-bound peptides that can be studied by PELDOR. The PELDOR time traces of the absorbed peptides indicate a fractal dimensionality $d = 2.1$, corresponding to absorption on a fairly planar surface. The distance distribution maximum around $r_{max} \sim 1.9$ nm for absorbed **Tri**^{1,8} was about the same as in methanol solution, with width $\Delta = 0.6$ nm, implying little change in their helical configuration.

In contrast, **Amp**^{3,13} has significantly different conformations in frozen solution and when absorbed. The $F(r)$ peak increases from 1.73 nm in methanol to 2.25 nm when absorbed on silica nanospheres and the width of the distance distribution increases greatly from 0.05 to 0.8 nm, Fig. 6.12. We can conclude that **Tri**^{1,8} remained a $2.2_7/3_{10}$ helix after absorption, but the **Amp**^{1,13} $F(r)$ is so broad that it is compatible with all helix conformations and even a β sheet, Fig. 6.1.

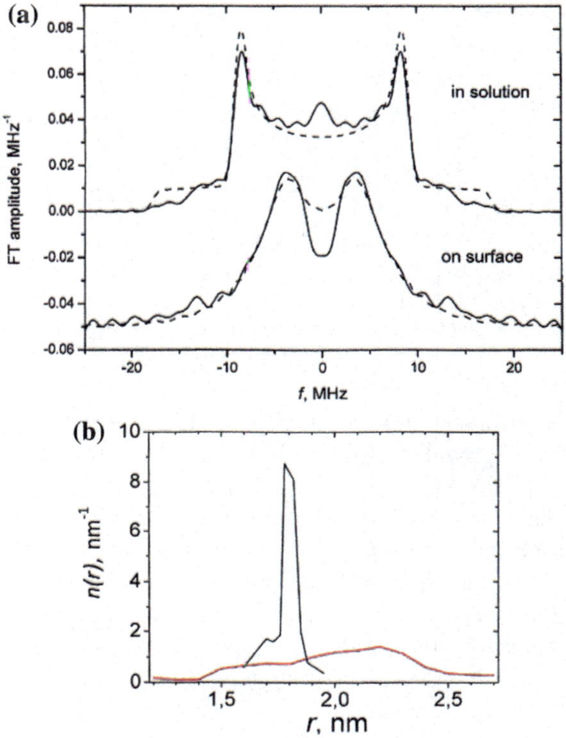

Fig. 6.12 **a** Cosine Fourier transforms (solid lines) and their simulations (*dashed lines*); and **b** $F(r)$ of the experimental $V_N(T)V_N(T)$ for $\mathbf{Amp}^{3,13}$ in frozen methanol and absorbed on silica nanospheres; the spectra for the absorbed sample in **a** are shifted downward by 0.05 MHz^{-1} for convenience. Reproduced from Syryamina et al. [73] with permission of Springer, copyright 2016

6.2 Spin-Labeled Alanine- and Proline-Based Peptides

The accuracy and precision, with which PELDOR techniques measure distances and peptide conformations, was verified with a series of well-characterized doubly spin-labeled peptides [74]. The peptides were based on a repeating sequence of four alanine and one lysine amino acid residues: $(A{\cdot}A{\cdot}A{\cdot}A{\cdot}K)_n{\cdot}A$ with $n = 4$ or 7. The sequence is written here with the standard, single-letter notation with lysine, alanine and cysteine amino acid residues indicated by K, A and C, respectively. Specific alanine residues were replaced by cysteine residues to which MTS **4-12** or SAT **4-15** spin labels were attached. The labeled peptides are indicated by the value of n and the positions of the labeled cysteines. For example, 4K(3, 7) indicates the sequence of **6-1**, Table 6.4.

A wide range of peptides from 4K(3, 7) up to 7K(3, 32) were studied. The CW-EPR lineshape at X-band and the 4pPELDOR signals were measured at 50 K in frozen aqueous solutions. The limited pulse bandwidth produced some distortion at the shorter inter-spin distances, so the variation of the PELDOR modulation depth with pump pulse length was used to help correct the measured distances [74]. The resulting distances were compared with MD calculations that assume an

Table 6.4 Amino acid sequences of alanine- and proline-rich peptides investigated by PELDOR spectroscopy; the label sites are indicated by bold characters[a]

	Sequence
6-1	A·A·**C**·A·K·A·**C**·A·A·K·A·A·A·A·K·A·A·A·A·K·A[b]
6-2	Ac-A·A·A·A·K·(**TOPP**)·A·K·A·A·A·A·A·K·A·A·K·A·(**TOPP**)·K·A·A·A·A-NH_2
6-3	Ac-A·A·A·A·K·(**C**$_{\mathbf{MTSSL}}$)·A·K·A·A·A·A·A·K·A·A·K·A·(**C**$_{\mathbf{MTSSL}}$)·K·A·A·A·A-NH_2
6-4	P·**P**·**H**·**G**·**G**·**G**·**W**·P·P·P·P·P·P·P·**P**·**H**·**G**·**G**·**G**·**W**
6-5	P·**P**·**H**·**G**·**G**·**G**·**W**·P·A·A·A·A·K·A·A·A·A·K·(**C**$_{\mathbf{MTSSL}}$)·A·A·A·A·K·A

[a]Standard, single-letter amino acid codes are used: *P* proline; *H* histidine; *G* glycine; *W* tryptophan; *A* alanine; *K* lysine; *C* cysteine
[b]This sequence is corresponds to 4K(3, 7) in the text

α-helical structure of the peptide. Eight peptides with $n = 4$ and one peptide with $n = 7$ provided EPR and PELDOR data at distances between 1.0 and 4.7 nm.

There is a strong correlation between the experimental and modeled mean distances, Fig. 6.13, which even extends to 4.7 nm for 7K(3, 32). PELDOR provided excellent agreement for distances greater than 2.0 nm. Corrections for pulse bandwidth limitations extended that lower bound to near 1.6 nm. A large part of the 'error bars' in Fig. 6.13 is due to the range of conformations that was present in liquid solution and that was preserved when the sample was frozen. Unfortunately, the figure does not indicate the distribution of distances from the MD modeling of the liquid solutions.

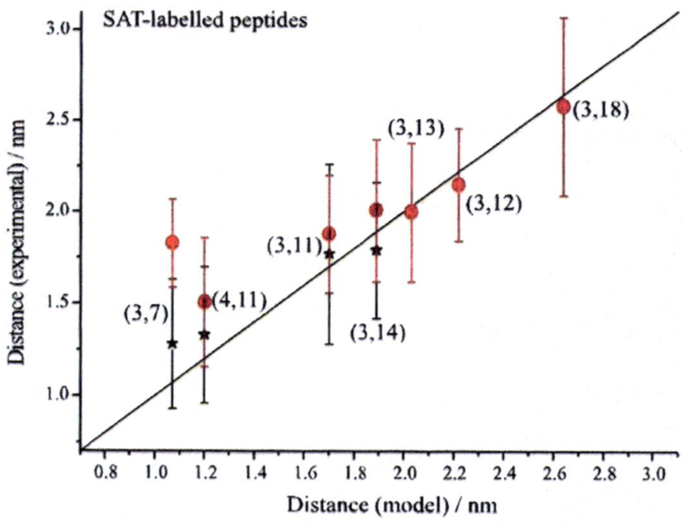

Fig. 6.13 Average distance and standard deviation from dipolar lineshape analysis (star) and PELDOR (red circle) for the SAT label on 4K(i, j) peptides; the model distance is the mean distance between spin labels from MD simulations. Reproduced from Banham et al. [74] with permissions of Elsevier, copyright 2008

The upper bound for CW-EPR measurements is in the range 1.5–1.7 nm. Exchange coupling and inhomogeneous EPR line broadening are a major source of errors in the overlap region near 1.5 nm where both CW-EPR and PELDOR are reliable. CW EPR spectroscopy and PELDOR techniques reliably measured distances in the alanine peptides spanning a wide range from 1.07 to over 4.6 nm [74].

The 4K family of alanine-based peptides are α-helical in aqueous solutions according to MD modeling and NMR and CD data but the helix is not rigid. The 7K peptide also seemed to be α-helical but PELDOR gave a somewhat shorter mean distance and broader distribution than expected for a completely rigid helix [74].

Two alanine-rich peptides, **6-2** and **6-3**, Table 6.4, labeled by TOPP **4-21** and Cys-MTSL **4-16** were synthesized and studied by X-band PELDOR in frozen solutions of TFE/EtOH/H_2O and TFE/EtOH/MeOH at 50 K. Tikhonov analysis of the PELDOR $V(T)$ of **6-2** indicated a narrow distance distribution centered at 2.80 nm having a width $\Delta = 0.26\,\text{nm}$. The distance is consistent with the estimate of 2.7 nm for two rigid spin labels linked to an α-helical peptide. A comparative study of peptide **6-3** gave an interspin distance of 2.26 nm, with a width as narrow as that of **6-2** [75]. MTSL labels usually give broad distance distributions, yet the observed distance between the MTSL is over 0.5 nm shorter than the distance between TOPP labels. These anomalies suggest that the MTSL labels may adopt a specific conformation in this peptide. This highlights a major advantage with the use of TOPP rather than MTSL: the interpretation of the distance is straightforward due to the rigid, reliable structure of the label. Unfortunately, the potential for orientation selection with the rigid TOPP labels was not exploited to determine their mutual orientation in the peptide.

Peptide chains have been used as linkers between paramagnetic ions and nitroxyls. Measurements of Cu^{2+}–Cu^{2+} and Cu^{2+}–nitroxyl distance distributions used PELDOR with a proline-based peptide **6-4** and an alanine-based peptide **6-5**, Table 6.4 [76, 77]. The proline-based peptide contains two well-characterized Cu^{2+} binding segments, P·H·G·G·G·W, separated by seven proline residues. Again, standard, single-letter amino acid codes are used: P, proline; H, histidine; G, glycine; W, tryptophan; A, alanine; K, lysine; C, cysteine. The P·H·G·G·G·W sequence in both peptides is a copper-binding sequence, while cysteine near one end of **6-5** was used to covalently attach the MTSL spin label. The P·H·G·G·G·W sites were occupied by Cu^{2+}.

PELDOR experiments were performed at several applied magnetic fields and resonance offsets to probe orientation selection effects on the Cu^{2+}-based PELDOR signal. Subtle orientation selection was observed from the PELDOR spectra of both peptides [76, 77]. A general theoretical model based on [78] was used to analyze the experimental data. Tikhonov regularization did not extract precise Cu^{2+}-based distance distributions. Instead, a full data analysis was required to obtain the distance distributions and relative orientations with a 3.0 nm mean Cu^{2+}–Cu^{2+} distance and a 2.7 nm mean Cu^{2+}-nitroxyl distance in the two peptides. These distances are consistent with structural models and with earlier measurements [79].

Constraints on the relative orientation between paramagnetic centers in these two model peptides were determined by examination of orientation selection effects. The analysis procedure is system independent, and therefore is applicable to more complicated biological systems.

References

1. Toniolo C, Crisma M, Formaggio F, Peggion C, Epand RF, Epand RM (2001) Lipopeptaibols, a novel family of membrane active, antimicrobial peptides. Cell Mol Life Sci 58(9):1179–1188. https://doi.org/10.1007/PI00000932
2. Peggion C, Formaggio F, Crisma M, Epand RF, Epand RM, Toniolo C (2003) Trichogin: a paradigm for lipopeptaibols. J Pept Sci 9(11–12):679–689. https://doi.org/10.1002/psc.500
3. Toniolo C, Crisma M, Formaggio F, Peggion C, Monaco V, Goulard C, Rebuffat S, Bodo B (1996) Effect of *N*-alpha-acyl chain length on the membrane-modifying properties of synthetic analogs of the lipopeptaibol trichogin GA IV. J Am Chem Soc 118(21):4952–4958. https://doi.org/10.1021/ja954081o
4. Rebuffat S, Goulard C, Bodo B, Roquebert MF (1999) The peptaibol antibiotics from *Trichoderma* soil fungi; structural diversity and membrane properties. In: Pandalai SG (ed) Recent research developments in organic & bioorganic chemistry, vol 3. Transworld Research Network, Trivandrum, pp 65–91
5. Fox RO, Richards FM (1982) A voltage-gated ion channel model inferred from the crystal-structure of alamethicin at 1.5 Å resolution. Nature 300(5890):325–330. https://doi.org/10.1038/300325a0
6. Karle IL, Flippenanderson J, Sukumar M, Balaram P (1987) Conformation of a 16-residue zervamicin-iia analog peptide containing 3 different structural features—3(10)-helix, alpha-helix, and beta-bend ribbon. Proc Natl Acad Sci USA 84(15):5087–5091. https://doi.org/10.1073/pnas.84.15.5087
7. Toniolo C, Peggion C, Crisma M, Formaggio F, Shui XQ, Eggleston DS (1994) Structure determination of racemic trichogin-a-IV using centrosymmetric crystals. Nat Struct Biol 1 (12):908–914. https://doi.org/10.1038/nsb1294-908
8. Crisma M, Monaco V, Formaggio F, Toniolo C, George C, Flippen-Anderson JL (1997) Crystallographic structure of a helical lipopeptaibol antibiotic analogue. Lett Pept Sci 4(4–6):213–218. https://doi.org/10.1023/A:1008874816982
9. Toniolo C, Brückner H (2007) Topical issue: peptaibiotics. Chem Biodivers 4(6):1021–1412. https://doi.org/10.1002/cbdv.200790093
10. Brückner H, Toniolo C (2013) Special issue: peptaibiotics II. Chem Biodivers 10(5):731–961. https://doi.org/10.1002/cbdv.201300139
11. Toniolo C, Brückner H (eds) (2009) Peptaibiotics fungal peptides containing α-dialkyl α-amino acids. Wiley-VCH, Weinheim
12. Toniolo C, Crisma M, Formaggio F, Peggion C (2001) Control of peptide conformation by the thorpe-ingold effect (C(alpha)-tetrasubstitution). Biopolymers 60(6):396–419. https://doi.org/10.1002/bip.10184
13. Karle IL, Balaram P (1990) Structural characteristics of alpha-helical peptide molecules containing aib residues. Biochem US 29(29):6747–6756. https://doi.org/10.1021/bi00481a001
14. Toniolo C, Crisma M, Formaggio F (1998) TOAC, a nitroxide spin-labeled, achiral C(alpha)-tetrasubstituted alpha-amino acid, is an excellent tool in material science and biochemistry. Biopolymers 47(2):153–158

15. Hanson P, Millhauser G, Formaggio F, Crisma M, Toniolo C (1996) ESR characterization of hexameric, helical peptides using double TOAC spin labeling. J Am Chem Soc 118 (32):7618–7625. https://doi.org/10.1021/ja961025u
16. Hanson P, Anderson DJ, Martinez G, Millhauser G, Formaggio F, Crisma M, Toniolo C, Vita C (1998) Electron spin resonance and structural analysis of water soluble, alanine-rich peptides incorporating TOAC. Mol Phys 95(5):957–966. https://doi.org/10.1080/002689798166576
17. Anderson DJ, Hanson P, McNulty J, Millhauser G, Monaco V, Formaggio F, Crisma M, Toniolo C (1999) Solution structures of TOAC-labeled trichogin GA IV peptides from allowed ($g \approx 2$) and half-field electron spin resonance. J Am Chem Soc 121(29):6919–6927. https://doi.org/10.1021/ja984255c
18. Bobone S, Roversi D, Giordano L, De Zotti M, Formaggio F, Toniolo C, Park Y, Stella L (2012) The lipid dependence of antimicrobial peptide activity is an unreliable experimental test for different pore models. Biochem US 51(51):10124–10126. https://doi.org/10.1021/bi3015086
19. Bobone S, Gerelli Y, De Zotti M, Bocchinfuso G, Farrotti A, Orioni B, Sebastiani F, Latter E, Penfold J, Senesi R, Formaggio F, Palleschi A, Toniolo C, Fragneto G, Stella L (2013) Membrane thickness and the mechanism of action of the short peptaibol trichogin GA IV. BBA-Biomem 1828(3):1013–1024. https://doi.org/10.1016/j.bbamem.2012.11.033
20. Tsvetkov YD, Milov AD, Maryasov AG (2008) Pulsed electron–electron double resonance (PELDOR) as EPR spectroscopy in nanometre range. Russ Chem Rev 77(6):487–520
21. Milov AD, Tsvetkov YD, Gorbunova EY, Mustaeva LG, Ovchinnikova TV, Raap J (2002) Self-aggregation properties of spin-labeled zervamicin IIA as studied by PELDOR spectroscopy. Biopolymers 64(6):328–336. https://doi.org/10.1002/bip.10208
22. Milov AD, Tsvetkov YD, Gorbunova EY, Mustaeva LG, Ovchinnikova TV, Handgraaf JW, Raap J (2007) Solvent effects on the secondary structure of the membrane-active zervamicin determined by PELDOR spectroscopy. Chem Biodivers 4(6):1243–1255. https://doi.org/10.1002/cbdv.200790107
23. Toniolo C, Benedetti E (1991) The polypeptide-3_{10}-helix. Trends Biochem Sci 16(9):350–353
24. Yasui SC, Keiderling TA, Bonora GM, Toniolo C (1986) Vibrational circular-dichroism of polypeptides. 5. A study of 3_{10}-helical-octapeptides. Biopolymers 25(1):79–89. https://doi.org/10.1002/bip.360250107
25. Crisma M, Formaggio F, Moretto A, Toniolo C (2006) Peptide helices based on alpha-amino acids. Biopolymers 84(1):3–12. https://doi.org/10.1002/bip.20357
26. Toniolo C (1980) Intramolecularly hydrogen-bonded peptide conformations. CRC Cr Rev Bioch Mol 9(1):1–44. https://doi.org/10.3109/10409238009105471
27. Crisma M, De Zotti M, Moretto A, Peggion C, Drouillat B, Wright K, Couty F, Toniolo C, Formaggio F (2015) Single and multiple peptide gamma-turns: literature survey and recent progress. New J Chem 39(5):3208–3216. https://doi.org/10.1039/c4nj01564a
28. Armen R, Alonso DOV, Daggett V (2003) The role of α-, 3_{10}-, and π-helix in helix $\rightarrow$ coil transitions. Protein Sci 12(6):1145–1157. https://doi.org/10.1110/ps02040103
29. Millhauser GL (1995) Views of helical peptides—a proposal for the position of 3_{10}-helix along the thermodynamic folding pathway. Biochem US 34(12):3873–3877. https://doi.org/10.1021/bi00012a001
30. Milov AD, Maryasov AG, Tsvetkov YD, Raap J (1999) Pulsed ELDOR in spin-labeled polypeptides. Chem Phys Lett 303(1–2):135–143. https://doi.org/10.1016/S0009-2614(99)00220-1
31. Milov AD, Maryasov AG, Samoilova RI, Tsvetkov YD, Raap J, Monaco V, Formaggio F, Crisma M, Toniolo C (2000) Pulsed double electron–electron resonance in spin-labeled polypeptides data on the secondary structure. Dokl Akad Nauk 370(2):265–268

32. Milov AD, Tsvetkov YD, Formaggio F, Crisma M, Toniolo C, Raap J (2001) The secondary structure of a membrane-modifying peptide in a supramolecular assembly studied by PELDOR and CW-ESR spectroscopies. J Am Chem Soc 123(16):3784–3789. https://doi.org/10.1021/ja0033990
33. Milov AD, Tsvetkov YD, Raap J, De Zotti M, Formaggio F, Toniolo C (2016) Conformation, self-aggregation, and membrane interaction of peptaibols as studied by pulsed electron double resonance spectroscopy. Biopolymers 106(1):6–24. https://doi.org/10.1002/bip.22713
34. Milov AD, Tsvetkov YD, Maryasov AG, Gobbo M, Prinzivalli C, De Zotti M, Formaggio F, Toniolo C (2012) Conformational properties of the spin-labeled tylopeptin B and heptaibin peptaibiotics based on PELDOR spectroscopy data. Appl Magn Reson 44(4):495–508. https://doi.org/10.1007/s00723-012-0402-1
35. Milov AD, Tsvetkov YD, De Zotti M, Prinzivalli C, Biondi B, Formaggio F, Toniolo C, Gobbo M (2013) Aggregation modes of the spin mono-labeled tylopeptin B and heptaibin peptaibiotics in frozen solutions of weak polarity as studied by PELDOR spectroscopy. J Struct Chem 54:S73–S85. https://doi.org/10.1134/S0022476613070056
36. Milov AD, Samoilova RI, Tsvetkov YD, De Zotti M, Toniolo C, Raap J (2008) PELDOR conformational analysis of bis-labeled alamethicin aggregated in phospholipid vesicles. J Phys Chem B 112(43):13469–13472. https://doi.org/10.1021/Jp8046714
37. Milov AD, Tsvetkov YD, Formaggio F, Oancea S, Toniolo C, Raap J (2003) Aggregation of spin labeled trichogin GA IV dimers: distance distribution between spin labels in frozen solutions by PELDOR data. J Phys Chem B 107(49):13719–13727. https://doi.org/10.1021/jp035057x
38. Milov AD, Tsvetkov YD, Formaggio F, Oancea S, Toniolo C, Raap J (2004) Solvent effect on the distance distribution between spin labels in aggregated spin labeled trichogin GA IV dimer peptides as studied by pulsed electron–electron double resonance. Phys Chem Chem Phys 6 (13):3596–3603. https://doi.org/10.1039/b313701e
39. Milov AD, Tsvetkov YD, Maryasov AG, Gobbo M, Prinzivalli C, De Zotti M, Formaggio F, Toniolo C (2013) Conformational properties of the spin-labeled tylopeptin B and heptaibin peptaibiotics based on PELDOR spectroscopy data. Appl Magn Reson 44(4):495–508. https://doi.org/10.1007/s00723-012-0402-1
40. Milov AD, Tsvetkov YD, Bortolus M, Maniero AL, Gobbo M, Toniolo C, Formaggio F (2014) Synthesis and conformational properties of a TOAC doubly spin-labeled analog of the medium-length, membrane active peptaibiotic ampullosporin A as revealed by CD, fluorescence, and EPR spectroscopies. Biopolymers 102(1):40–48. https://doi.org/10.1002/bip.22362
41. Rabenstein MD, Shin YK (1995) Determination of the distance between 2 spin labels attached to a macromolecule. Proc Natl Acad Sci USA 92(18):8239–8243. https://doi.org/10.1073/pnas.92.18.8239
42. Bortolus M, Tombolato F, Tessari I, Bisaglia M, Mammi S, Bubacco L, Ferrarini A, Maniero AL (2008) Broken helix in vesicle and micelle-bound alpha-synuclein: insights from site-directed spin labeling-EPR experiments and MD simulations. J Am Chem Soc 130 (21):6690–6691. https://doi.org/10.1021/jaB010429
43. Karle IL (1994) Diffraction studies of model and natural helical peptides. In: White SH (ed) Membrane protein structure: experimental approaches. Springer, New York, pp 355–380. https://doi.org/10.1007/978-1-4614-7515-6
44. Balashova TA, Shenkarev ZO, Tagaev AA, Ovchinnikova TV, Raap J, Arseniev AS (2000) NMR structure of the channel-former zervamicin IIB in isotropic solvents. FEBS Lett 466(2–3):333–336. https://doi.org/10.1016/S0014-5793(99)01707-X
45. Peggion C, Coin I, Toniolo C (2004) Total synthesis in solution of alamethicin F50/5 by an easily tunable segment condensation approach. Biopolymers 76(6):485–493. https://doi.org/10.1002/bip.20161
46. Crisma M, Peggion C, Baldini C, MacLean EJ, Vedovato N, Rispoli G, Toniolo C (2007) Crystal structure of a spin-labeled, channel-forming alamethicin analogue. Angew Chem Int Edit 46(12):2047–2050. https://doi.org/10.1002/anie.200604417

47. Milov AD, Tsvetkov YD, Formaggio F, Crisma M, Toniolo C, Raap J (2000) Self-assembling properties of membrane-modifying peptides studied by PELDOR and CW-ESR spectroscopies. J Am Chem Soc 122(16):3843–3848. https://doi.org/10.1021/ja993870t
48. Milov AD, Tsvetkov YD, Raap J (2000) Aggregation of trichogin analogs in weakly polar solvents: PELDOR and ESR studies. Appl Magn Reson 19(2):215–226. https://doi.org/10.1007/Bf03162276
49. Milov AD, Tsvetkov YD, Formaggio F, Crisma M, Toniolo C, Raap J (2003) Self-assembling and membrane modifying properties of a lipopeptaibol studied by CW-ESR and PELDOR spectroscopies. J Pept Sci 9(11–12):690–700. https://doi.org/10.1002/psc.513
50. Milov AD, Tsvetkov YD, Formaggio F, Crisma M, Toniolo C, Millhauser GL, Raap J (2001) Self-assembling properties of a membrane-modifying lipopeptaibol in weakly polar solvents, studied by CW ESR. J Phys Chem B 105(45):11206–11213. https://doi.org/10.1021/Jp011948y
51. Milov AD, Samoilova MI, Tsvetkov YD, Jost M, Peggion C, Formaggio F, Crisma M, Toniolo C, Handgraaf JW, Raap J (2007) Supramolecular structure of self-assembling alamethicin analog studied by ESR and PELDOR. Chem Biodivers 4(6):1275–1298. https://doi.org/10.1002/cbdv.200790110
52. Milov AD, Samoilova RI, Tsvetkov YD, Peggion C, Formaggio F, Toniolo C, Raap J (2006) Aggregation of spin-labeled alamethicin in low-polarity solutions as studied by PELDOR spectroscopy. Dokl Phys Chem 406:21–25. https://doi.org/10.1134/S0012501606010064
53. Marsh D (1989) Experimental methods in spin-label spectral analysis. In: Berliner LJ, Reuben J (eds) Spin labeling, theory and applications. Biological magnetic resonance, vol 8. Plenum Press, New York, pp 255–303. https://doi.org/10.1007/978-1-4613-0743-3
54. Stella L, Mazzuca C, Venanzi M, Palleschi A, Didone M, Formaggio F, Toniolo C, Pispisa B (2004) Aggregation and water-membrane partition as major determinants of the activity of the antibiotic peptide trichogin GA IV. Biophys J 86(2):936–945
55. Kropacheva TN, Raap J (2002) Ion transport across a phospholipid membrane mediated by the peptide trichogin GA IV. Biochim Biophys Acta (BBA)—Biomemb 1567:193–203. https://doi.org/10.1016/s0005-2736(02)00616-8
56. Milov AD, Samoilova RI, Tsvetkov YD, Formaggio F, Toniolo C, Raap J (2005) Membrane-peptide interaction studied by PELDOR and CW ESR: peptide conformations and cholesterol effect on the spatial peptide distribution in the membrane. Appl Magn Reson 29(4):703–716. https://doi.org/10.1007/Bf03166345
57. Salnikov ES, Erilov DA, Milov AD, Tsvetkov YD, Peggion C, Formaggio F, Toniolo C, Raap J, Dzuba SA (2006) Location and aggregation of the spin-labeled peptide trichogin GA IV in a phospholipid membrane as revealed by pulsed EPR. Biophys J 91(4):1532–1540. https://doi.org/10.1529/biophysj.105.075887
58. Milov AD, Erilov DA, Salnikov ES, Tsvetkov YD, Formaggio F, Toniolo C, Raap J (2005) Structure and spatial distribution of the spin-labelled lipopeptide trichogin GA IV in a phospholipid membrane studied by pulsed electron–electron double resonance (PELDOR). Phys Chem Chem Phys 7(8):1794–1799. https://doi.org/10.1039/b418414a
59. Milov AD, Samoilova RI, Tsvetkov YD, Formaggio F, Toniolo C, Raap J (2007) Self-aggregation of spin-labeled alamethicin in ePC vesicles studied by pulsed electron–electron double resonance. J Am Chem Soc 129(30):9260–9261. https://doi.org/10.1021/ja072851d
60. Bartucci R, Guzzi R, De Zotti M, Toniolo C, Sportelli L, Marsh D (2008) Backbone dynamics of alamethicin bound to lipid membranes: spin-echo electron paramagnetic resonance of TOAC-spin labels. Biophys J 94(7):2698–2705. https://doi.org/10.1529/biophysj.107.115287
61. Salnikov ES, De Zotti M, Formaggio F, Li X, Toniolo C, O'Neil JDJ, Raap J, Dzuba SA, Bechinger B (2009) Alamethicin topology in phospholipid membranes by oriented solid-state NMR and EPR spectroscopies: a comparison. J Phys Chem B 113(10):3034–3042. https://doi.org/10.1021/jp8101805

62. Milov AD, Samoilova RI, Tsvetkov YD, De Zotti M, Formaggio F, Toniolo C, Handgraaf JW, Raap J (2009) Structure of self-aggregated alamethicin in ePC membranes detected by pulsed electron–electron double resonance and electron spin echo envelope modulation spectroscopies. Biophys J 96(8):3197–3209. https://doi.org/10.1016/j.bpj.2009.01.026
63. North CL, BarrangerMathys M, Cafiso DS (1995) Membrane orientation of the N-terminal segment of alamethicin determined by solid-state N-15 NMR. Biophys J 69(6):2392–2397
64. Salnikov ES, Friedrich H, Li X, Bertani P, Reissmann S, Hertweck C, O'Neil JDJ, Raap J, Bechinger B (2009) Structure and alignment of the membrane-associated peptaibols ampullosporin A and alamethicin by oriented N-15 and P-31 solid-state NMR spectroscopy. Biophys J 96(1):86–100. https://doi.org/10.1529/biophysj.108.136242
65. Marsh D, Jost M, Peggion C, Toniolo C (2007) Lipid chain-length dependence for incorporation of alamethicin in membranes: electron paramagnetic resonance studies on TOAC-spin labeled analogs. Biophys J 92(11):4002–4011. https://doi.org/10.1529/biophysj.107.104026
66. Parsegian VA, Fuller N, Rand RP (1979) Measured work of deformation and repulsion of lecithin bilayers. Proc Natl Acad Sci USA 76(6):2750–2754. https://doi.org/10.1073/pnas.76.6.2750
67. Murzyn K, Rog T, Blicharski W, Dutka M, Pyka J, Szytula S, Froncisz W (2006) Influence of the disulfide bond configuration on the dynamics of the spin label attached to cytochrome *c*. Proteins-Struct Funct Bioinform 62(4):1088–1100. https://doi.org/10.1002/prot.20838
68. Spaar A, Munster C, Salditt T (2004) Conformation of peptides in lipid membranes studied by X-ray grazing incidence scattering. Biophys J 87(1):396–407. https://doi.org/10.1529/biophysj.104.040667
69. Reginsson GW, Schiemann O (2011) Studying bimolecular complexes with pulsed electron–electron double resonance spectroscopy. Biochem Soc T 39:128–139. https://doi.org/10.1042/Bst0390128
70. Goobes G, Stayton PS, Drobny GP (2007) Solid-state NMR studies of molecular recognition at protein-mineral interfaces. Prog Nucl Mag Res Sp 50(2–3):71–85. https://doi.org/10.1016/j.pnmrs.2006.11.002
71. Mirau PA, Naik RR, Gehring P (2011) Structure of peptides on metal oxide surfaces probed by NMR. J Am Chem Soc 133(45):18243–18248. https://doi.org/10.1021/ja205454t
72. Milov AD, Samoilova RI, Tsvetkov YD, Peggion C, Formaggio F, Toniolo C (2014) Peptides on the surface. PELDOR data for spin-labeled alamethicin F50/5 analogues on organic sorbent. J Phys Chem B 118(25):7085–7090. https://doi.org/10.1021/jp503691n
73. Syryamina VN, Samoilova RI, Tsvetkov YD, Ischenko AV, De Zotti M, Gobbo M, Toniolo C, Formaggio F, Dzuba SA (2016) Peptides on the surface: spin-label EPR and PELDOR study of adsorption of the antimicrobial peptides trichogin GA IV and ampullosporin A on the silica nanoparticles. Appl Magn Reson 47(3):309–320. https://doi.org/10.1007/s00723-015-0745-5
74. Banham JE, Baker CM, Ceola S, Day IJ, Grant GH, Groenen EJJ, Rodgers CT, Jeschke G, Timmel CR (2008) Distance measurements in the borderline region of applicability of CW EPR and DEER: a model study on a homologous series of spin-labelled peptides. J Magn Reson 191(2):202–218. https://doi.org/10.1016/j.jmr.2007.11.023
75. Stoller S, Sicoli G, Baranova TY, Bennati M, Diederichsen U (2011) TOPP: a novel nitroxide-labeled amino acid for EPR distance measurements. Angew Chem Int Edit 50(41):9743–9746
76. Yang ZY, Kise D, Saxena S (2010) An approach towards the measurement of nanometer range distances based on Cu^{2+} ions and ESR. J Phys Chem B 114(18):6165–6174. https://doi.org/10.1021/jp911637s
77. Yang ZY, Ji M, Saxena S (2010) Practical aspects of copper ion-based double electron electron resonance distance measurements. Appl Magn Reson 39(4):487–500. https://doi.org/10.1007/s00723-010-0181-5

78. Maryasov AG, Tsvetkov YD, Raap J (1998) Weakly coupled radical pairs in solids: ELDOR in ESE structure studies. Appl Magn Reson 14(1):101–113. https://doi.org/10.1007/Bf03162010
79. Jun S, Becker JS, Yonkunas M, Coalson R, Saxena S (2006) Unfolding of alanine-based peptides using electron spin resonance distance measurements. Biochem US 45(38):11666–11673. https://doi.org/10.1021/bi061195b

Chapter 7
Structural Studies of Nucleic Acids

Abstract Nucleic acids have a very rich range of structures that are important in many biological contexts. PELDOR or DEER spectroscopy has provided a unique glimpse at the structures that form in solution and guide the response of the cellular machinery.

DNA and RNA nucleic acids have backbones consisting of chains of alternating sugar and phosphate groups linked by ester bonds. A nitrogenous base is attached to each sugar. The bases interact with each other and tend to stabilize DNA in a double helix Fig. 7.1, while RNA has a richer range of structures.

The bases commonly found in DNA are adenine (A), guanine (G), cytosine (C) and thymine (T). RNA is similar, but contains uracil (U) instead of thymine, and the sugar is ribose instead of deoxyribose. Each nitrogenous base can form stable hydrogen bonds with another base to stabilize the three-dimensional structure of the nucleic acid. Thymine pairs with adenine, while guanine pairs with cytosine. In RNA, uracil pairs with adenine.

Nucleic acids are the carriers of the genetic code in biological systems, but also have important regulatory, structural and catalytic roles. Cells modify particular bases, often by methylation, to cause major and minor structural changes that they use to regulate nucleic acid functions. Cells also need to recognize the structural changes in DNA that is damaged, e.g., by radiation or oxidative stress, so that the damage can be repaired or destroyed. Structural information about intact and

© Springer Nature Switzerland AG 2019

Y. D. Tsvetkov et al., *Pulsed Electron–Electron Double Resonance*,
https://doi.org/10.1007/978-3-030-05372-7_7

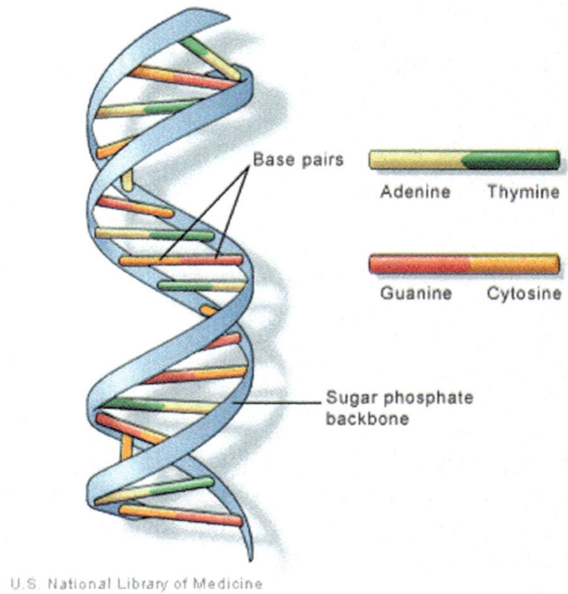

Fig. 7.1 Schematic representation of DNA structure. Courtesy of the National Library of Medicine

damaged DNA, RNA and their complexes in various biological systems is a primary goal of many physical studies, including PELDOR, in biology.

The success of PELDOR research in this area is a direct result of the effective methods developed for site-directed spin labeling, Sect. 4.3. PELDOR studies of DNA and RNA have investigated several specific biochemical systems. However, this work is only briefly discussed in reviews [1–7] and only in combination with other applications of PELDOR. This Chapter covers the application of PELDOR to DNA and RNA.

7.1 Linear Duplexes of Nucleic Acids

7.1.1 Duplex Formation

The structures of 12-bp (base pair) duplex **RNA1**, with all bases paired, and 15-bp duplex **RNA2**, with only eight paired bases at the center and unpaired bases at both ends, were investigated [8]. TEMPO labels **4-19** were attached to the amino groups of 2-amino-ribose in 2'-amino uridine **U** residues via reaction with TEMPO isothiocyanates.

5'—G-C-U-G-A-**U**-A-U-C-A-G-C—3'
3'—C-G-A-C-U-A-**U**-A-G-U-C-G—5'

RNA1

5'—G-G-C-C-A-G-A-**U**-C-U-G-A-G-C-G—3'
3'—G-C-G-A-G-U-C-**U**-A-G-A-C-C-G-G—5'

RNA2

U =

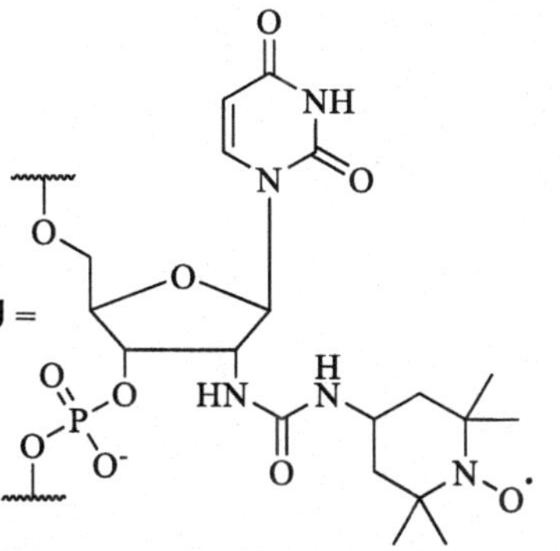

A modulated 3pPELDOR time trace *V(T)* was seen in aqueous buffer only for **RNA1**. The distance between spin labels $r = 3.5 \pm 0.2$ nm was determined by Fourier analysis. The *V*(*T*) for **RNA2** showed only an exponential decay, which indicated a uniform volume distribution of spin labels. This means that no duplexes formed between spin-labeled **RNA2** oligonucleotides at 0.3 mM in aqueous buffer: 0.1 M NaCl, 0.01 M Na-phosphate, 0.1 mM Na_2EDTA, pH 7.2. Both RNA sequences are palindromic, indicating that the duplex sequence would be identical when read from either end of the duplex. Unfortunately, this also means that intramolecular hairpin structures with only one spin label can form instead of the desired bimolecular duplex. Na-phosphate is a notoriously poor choice of buffer for frozen solutions because it can undergo extreme shifts in pH as it freezes [9].

A series of 27-bp DNA duplexes was studied with spin labels attached at 2′-amino uridine residues in different strands by reaction with TEMPO isocyanate. The DNA duplexes had interspin distances of 9, 12, 15, 18 or 21 bp:

5'—G-C-**U**-C-G-**U**-C-A-**U**-G-G-**U**-C-C-**U**-C-A-T-T-G-T-A-C-T-A-G-C—3'
3'—C-G-A-G-C-A-G-T-A-C-C-A-G-G-A-G-T-A-A-C-A-T-G-A-**U**-C-G—5'

with the spin-labeled nucleotide sites shown in bold [10]. Five different doubly spin-labeled duplexes were synthesized. The spin label pairs with distances ranging from 2.8 to 6.8 nm are shown schematically in Fig. 7.2.

The distance distribution *F*(*r*) of individual duplexes gave distinct peaks at distances corresponding to the expected interspin distances. More interestingly, mixtures of two duplexes have *F(r)* that show the distances for both duplexes in the mixture, with amplitudes reflecting the relative concentrations, Fig. 7.3.

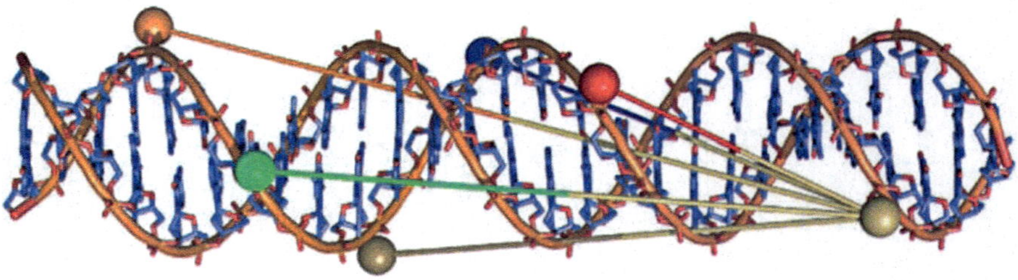

Fig. 7.2 Molecular model showing the positions of spin label pairs for distance measurements. Reprinted from Ward et al. [10] with permission of Wiley and Sons, copyright 2007

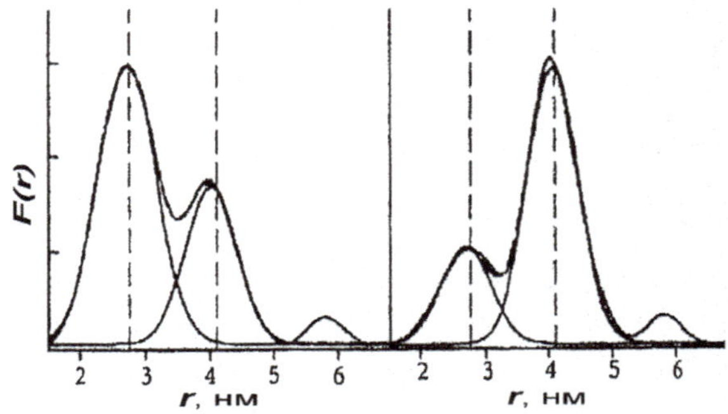

Fig. 7.3 Mixtures of spin-labeled DNA duplexes: $F(r)$ are shown for mixtures of labeled DNA duplexes with distances of 2.8 nm and 4.1 nm, **Left**: in a 3/1 ratio; **Right**: in a 1/3 ratio. Reprinted from Ward et al. [10] with permission of Wiley and Sons, copyright 2007

Deconvolution of $F(r)$ for the mixture using a Gaussian curve for each duplex yields the average distance in each duplex and its concentration in the mixture with good accuracy [10].

7.1.2 Duplex Conformation

TPA **4-18** spin labels were introduced via reaction B in Sect. 4.3.3 into 2′-deoxy-uridine dU residues of five different DNA duplexes and distances between the labels were determined using 4pPELDOR [11]. The number of bases between the labels, n, was different: $n = 0, 2, 8, 10, 12$, for **DNA1**–**DNA5**, respectively. For instance, **DNA1** and **DNA5**, with the spin-labeled dU residues indicated here as **U**, are:

3'—G-C-G-A-**U**-A-C-A-T-G-C-G—5'
5'—C-G-C-T-A-**U**-G-T-A-C-G-C—3'
DNA1

3'—G-C-**U**-G-A-C-T-A-T-A-T-A-G-T-C-A-G-C—5'
5'—C-G-A-C-T-G-A-T-A-T-A-T-C-A-G-**U**-C-G—3'
DNA5

The duplexes were investigated at 35 K in aqueous buffered solutions with 20% ethylene glycol added to form a good glass. The modulation of the PELDOR *V*(*T*) was recorded for all DNA molecules, Fig. 7.4. The modulation period increased as the distance between the spin labels increased. The Fourier spectra in all the cases had the shape of Pake doublets, Fig. 7.5. The features at 7.4 and 14.8 MHz in this doublet correspond to the parallel $\theta = 0°$ and perpendicular $\theta = 90°$ orientations of the vector connecting the spin labels relative to the external magnetic field direction. The distance between the labels is r = 1.92 nm from Eqs. 1.1 and 1.2 with the exchange integral $J = 0$. The distances for **DNA2–DNA5** were 2.33, 3.47, 4.48, and 5.25 nm, respectively. These distances, with their uncertainties, are in excellent agreement with values from MD simulations for the B-conformation of the duplex helix, Fig. 7.6 [11]. The correlation coefficient for the two sets of data is 0.997, supporting existence of the B-conformation in frozen aqueous solutions [11]. A detailed comparison of PELDOR and FRET [11] measurements on these oligomers demonstrated that these methods complement each other well.

Six RNA duplexes labeled with TPA **4-18** were synthesized and their *V*(*T*) had sufficiently deep modulation to calculate the *F*(*r*) [12]. The distances between spins in the duplexes range from 1.93 ± 0.12 to 3.87 ± 0.13 nm, depending on the number of base pairs between labels. Comparison of these experimental results in RNA [12] with the results just discussed for the same spin label in DNA [11] indicate that distances between labels on different helices are different in DNA and RNA. For example, for labels separated by 10 bp, the distance between them is 4.48 ± 0.50 nm in DNA but 3.87 ± 0.13 nm RNA [12]. The difference is too large to be accounted for by measurement error; but it does correspond to two different nucleic acid conformations: the A-form in RNA and the more-stretched B-form in DNA. The experimental distances are in good agreement with MD simulations: the correlation coefficient was 0.976 [12]. The DNA and RNA duplexes maintain their conformations in frozen aqueous phosphate buffer solutions with ethylene glycol.

7.1.3 Labeled Phosphate Groups

TPA **4-18** labels have been attached via a methylene linker to phosphorothioate groups at specific positions along the sugar-phosphate backbone rather than onto the nitrogenous base [13]. Eight interspin distances from 2.56 to 3.88 nm were

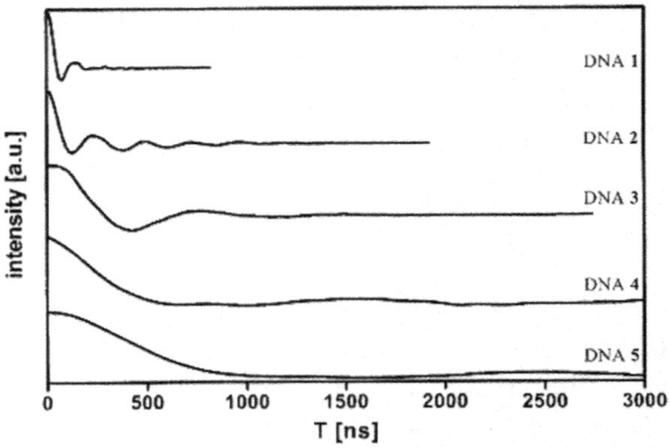

Fig. 7.4 PELDOR time traces $V_{INTRA}(T)$ for spin-labeled **DNA1-DNA5**. Reprinted from Schiemann et al. [11] with permission of American Chemical Society, copyright 2004

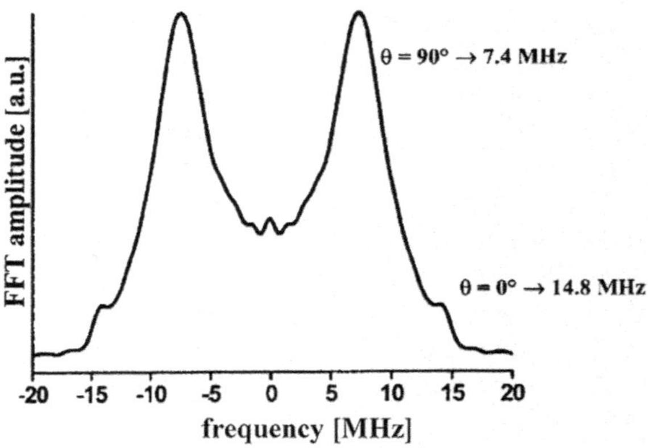

Fig. 7.5 Fourier spectrum of V_{INTRA} for **DNA1** showing a broadened Pake pattern. Reprinted from Schiemann et al. [11] with permission of American Chemical Society, copyright 2004

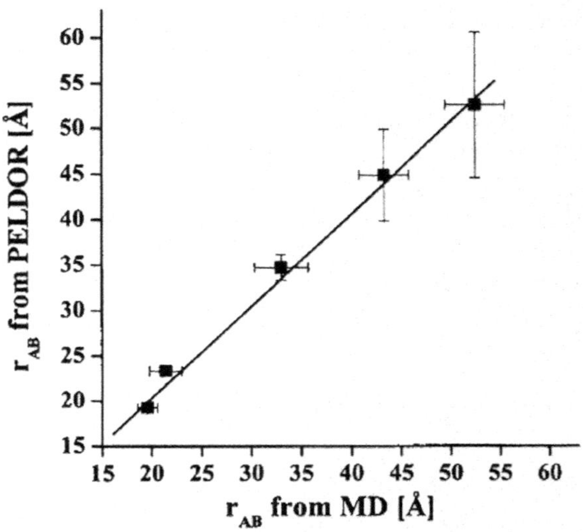

Fig. 7.6 Correlation of distances r_{AB} obtained by PELDOR experiment and MD calculations. Reprinted from Schiemann et al. [11] with permission of American Chemical Society, copyright 2004

determined in a set of 12-bp DNA duplexes from the maxima of peaks in the PELDOR *F*(*r*). This method of labeling readily scales to long polynucleotides [13]. An example of a 68-bp DNA fragment with labels on opposite strands 9 nucleotides from one end of the duplex is given. The 2.52 nm distance between spins that is measured by PELDOR for this duplex agreed with the 2.5 nm distance from MD simulations.

The distance from PELDOR was compared with that from NMR, giving an excellent correlation of $R^2 = 0.98$ [13]. This labeling strategy was proposed for structural studies of DNA and RNA complexes with protein.

A similar investigation introduced spin labels into phosphorothioate groups of RNA [14]. Six sets of interspin distances ranging from 2.5 to 4.72 nm were compared to X-ray structures. A strong correlation between these measurements was found with $R^2 = 0.97$. The introduction of the label does not seem to significantly alter the nucleic acid structure.

7.1.4 End Labels

It is rather easy to attach a label to the 5′ end of a nucleic acid duplex of any length. Larger labels can be accommodated at the end of a duplex than along the side of a duplex. This opens up the use of Trityl **T 4-2** spin labels, with a narrower spectrum and slower relaxation than nitroxyls, for EPR distance measurements. The use of trityl labels was tested in a series of model DNA duplexes **D1**, **D2** end-labeled with either **T** or nitroxyl **NIT** [15]. The spin labels were attached to the 5’ end by two different linkers, a short $-NH_2$- and a longer linker known as PIP:

D1 5'—C-A-C-G-C-C-G-C-T-G—3'
3'—G-T-G-C-G-G-C-G-A-C—5'

D2 5'—C-G-C-G-A-T-C-G-C-G—3'
3'—G-C-G-C-T-A-G-C-G-C—5'

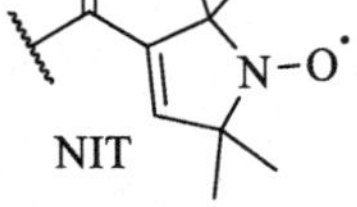

NIT

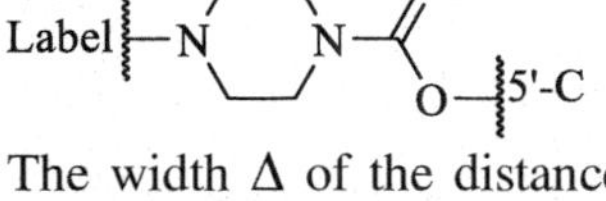

The width Δ of the distance distributions, obtained by Q-band 4pPELDOR at 80 K, strongly depends on the type of label. Replacement of both **NIT** by **T** in the same duplex sharpens the distance distribution, Table 7.1. Duplexes with one **NIT** and one **T**, sometimes called orthogonal labeling, have a less pronounced sharpening, but a significant gain in sensitivity due to more efficient pumping of the narrow EPR spectrum of **T**, Fig. 7.7. The width of the distance distribution Δ with

Table 7.1 Distances *r* and distance distribution widths Δ in DNA duplexes **D1**, **D2** labeled by trityls **T** and nitroxyls **NIT** [15]

DNA Complex	Distance, *r* (nm)	½ Δ (nm)*
D1-PIP-**T/T**	4.54	0.20
D1-PIP-**NIT/NIT**	4.24	0.62
D1-NH-**NIT/NIT**	3.91	0.37
D2-NH-**T/T**	4.49	0.25
D2-PIP-**T/T**	4.50	0.18
D2-NH-**NIT/NIT**	4.05	0.36

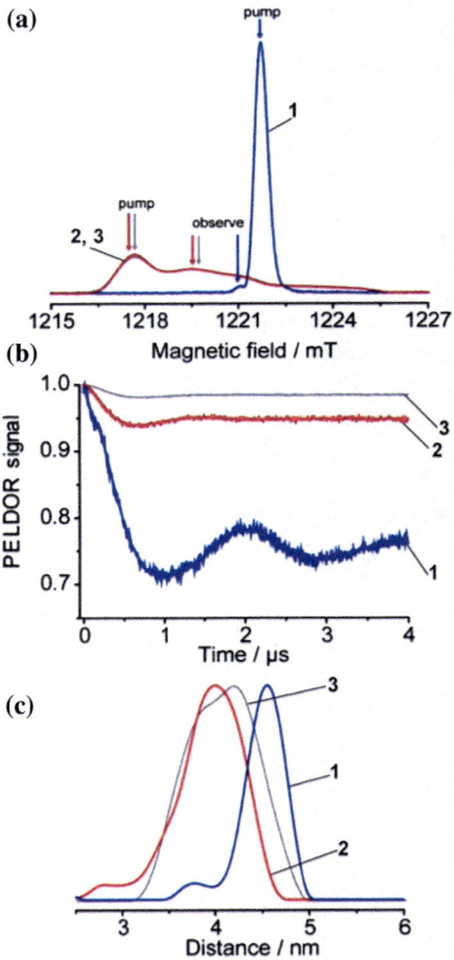

Fig. 7.7 **a** Echo-detected EPR spectra of duplexes: *1*: **D2**-NH-**T/T**, *2*: **D1**-NH-**NIT/NIT** and *3*: **D2**-NH-**NIT/NIT**, spectral positions of pump and observe pulses are indicated for each duplex; **b** PELDOR *V*(*T*) for each duplex after removal of relaxation background; **c** *F* (*r*) obtained for each duplex. Reproduced from Shevelev et al. [15] with permission of American Chemical Society, copyright 2015

NIT/NIT labels does depend on the linker: the shorter amine-based linker improves Δ by a factor of two. On the other hand, the linker length has no effect on Δ with **T/T** labels. MD calculations indicate greater conformational disorder of the **NIT**

labels compared to **T**, thus rationalizing the experimentally observed trends. In fact, the MD calculations show that hydrophobic areas on the **T** label cap the stack of bases at the end of the duplex [15].

Labeling with **T** gives narrower spin-spin distance distributions and potentially more precise distances between labeling sites than does more traditional nitroxyl labels. However, the distances in **NIT/NIT** pairs with the short linker are actually nearer to the length of ten-bp B-helix DNA. The ~1 nm diameter of **T**-labels must be considered in distance measurements [15].

The ten-bp **D1**-PIP-**T/T** and **D2**-PIP-**T/T** duplexes were measured by DQC at 310 K and at 80 K by PELDOR, giving the same distance, ~4.6 nm [16]. The DNA was immobilized on NucleosilDMA sorbent, which prevented movement of the **T** label.

Gd(III) [17–19] or Cu(II) [20] complexes have been recently suggested for use as labels, Sects. 4.2 and 4.4. These labels are typically characterized by a rather complex EPR spectrum in polyoriented systems. However, for Gd(III), the $-1/2 \rightarrow +1/2$ transition, dominates the spectrum at Ka-band for temperatures of ~10 K and is used in PELDOR experiments.

The structure of a 14-bp duplex was investigated using Gd(III) complexes **Gd538** and **Gd595** as labels incorporated at the terminal thymidine molecules using "click chemistry" [19]. The 4pPELDOR measurements found a distance of 5.9 ± 1.2 nm between the ions in these DNA duplexes, while the width of the distribution was ~2 nm. The use of Gd(III) labels might increase the range of PELDOR-measured distances to ~10 nm [19], which is significant for some complex biomolecules. The relatively large distance, 1.2–1.5 nm, between the ions and the position of attachment to the macromolecule results in a wide distance distribution and a decrease in the measurement accuracy due to the mobility of these labels. This is an obvious drawback of these labels in comparison to nitroxyl labels. A number of features of the PELDOR analysis methods for Gd(III) and Cu (II) have been thoroughly examined [19, 20].

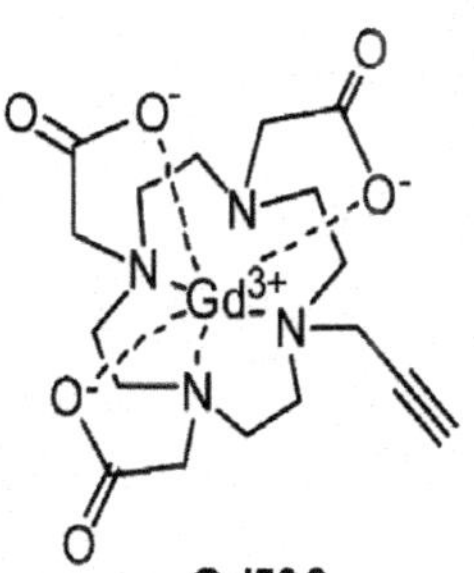

Gd538

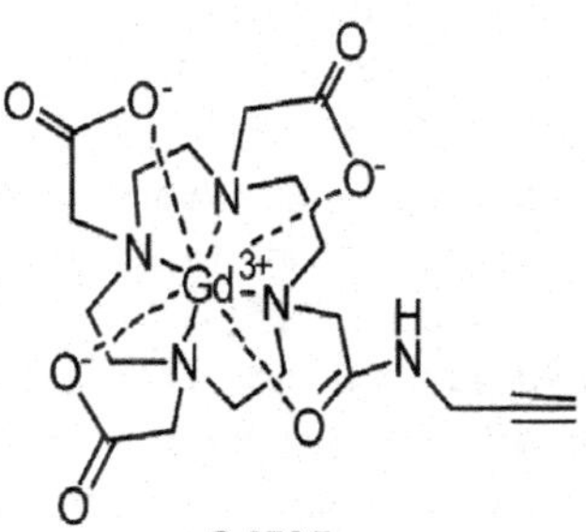

Gd595

```
(Gd)—T-C-T-A-C-T-G-C-T-T-T-A-G-A—3'
     3'—A-G-A-T-G-A-C-G-A-A-A-T-C-T—(Gd)
```

7.1.5 Conformational Changes

The studies just described show that spin labeling of linear DNA and RNA duplexes allow a rather consistent determination, to ~1%, of the distance between the spin-labeled nucleotides.

The strong correlation between PELDOR and MD results is very significant. The MD simulations are typically carried out for room temperatures and aqueous solutions, while the PELDOR measurements were made with rapidly-frozen glassy solutions. The conformations of DNA and RNA molecules at room temperature seem to persist even through flash freezing of the solutions. This suggests that PELDOR will give relevant structural information for nucleic acids in other environments, with novel interactions, and during reactions.

Changes in the distance between nucleotides during the transition from the B- to the A-form of DNA have been studied using PELDOR [21]. A series of spin-labeled DNA duplexes were investigated:

5'—G-A-C-C-C-T-G-C-A-T-C-T-A-**G**-G-A-T-**G**-C-C—3'
3'—C-T-G-**G**-G-A-C-G-T-A-G-A-T-C-C-T-A-C-**G**-**G**—5'

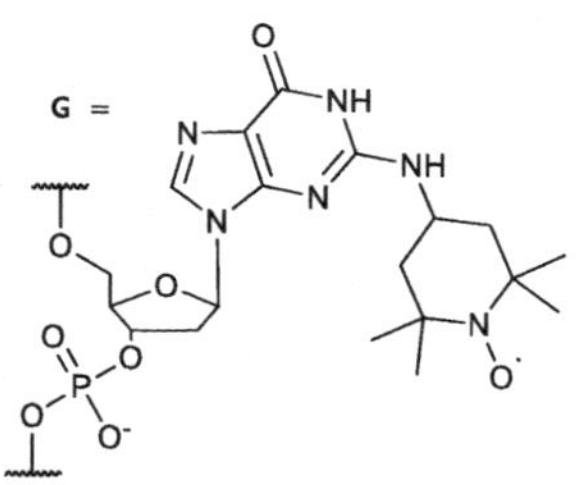

where the sites of spin labels are shown in bold.

The 4-amino-TEMPO **4-19** label was attached to the N2 atom of specific guanine residues. Labels were attached either to the same helix at positions denoted **(4;19)** and **(4;20)**, or to different helices at positions denoted **(4;14′)** or **(4;18′)**. The transition between the B- and A-forms of DNA occurs in polar media. PELDOR measurements were made at 60–70 K in aqueous buffer with 10 vol. % glycerol added as a cryoprotectant [21].

Addition of trifluoroethanol, TFE, stimulated the B → A transition. The distance spectrum changed with the volumetric content of TFE, Fig. 7.8. The longer B-form converts into the more compact A-form as the TFE concentration exceeds 70%. The measured interspin distances in the A- and B-forms agree well with distances between the oxygen atoms of the >NO groups of the spin labels from MD simulations, Table 7.2 [21].

The ability of PELDOR to determine distances in the nanometer range is considerably better than that of any other method, such as CW EPR or circular

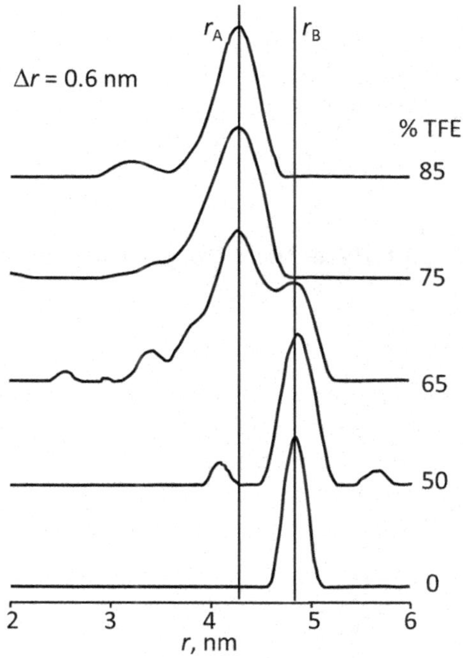

Fig. 7.8 The measured F(r) as TFE is added to an aqueous solution of **(4;18′)**-labeled DNA duplex. Reprinted from Sicoli et al. [21] with permission of Wiley and Sons, copyright 2008

Table 7.2 Experimental and calculated distances in nm between the spin labels for the A- and B-forms of DNA [21]

DNA duplex	PELDOR distance	Calculated O–O distance, B-form	PELDOR distance	A-form, calculated O–O distance
(4;20)	5.6 ± 0.2	5.6 ± 0.3	4.8 ± 0.2	4.5 ± 0.4
(4;19)	5.1 ± 0.2	5.1 ± 0.3	4.6 ± 0.3	4.4 ± 0.4
(4;18′)	4.9 ± 0.2	4.8 ± 0.4	4.3 ± 0.3	4.2 ± 0.4
(4;14′)	3.2 ± 0.2	3.6 ± 0.3	2.8 ± 0.3	3.3 ± 0.3

dichroism [21]. This enabled the investigation of the transitions between the A- and B-forms of RNA and DNA in various molecular environments and polarity.

The structure of the nucleic acid can also be altered by the spin label. MD simulations show that a TPA **4-18** spin label in the major groove of DNA or RNA changes the mutual orientation of the base pairs in the molecule [22]. This effect is less significant for labels located in the minor groove. Nonetheless, conformational changes do occur when DNA or RNA is labeled and should be taken into account when interpreting PELDOR results and when planning PELDOR experiments. The impact of the label was shown in four hybrid DNA/RNA-duplexes. Distances between spin labels have been measured with TPA spin labels attached to the nitrogenous bases so that they were oriented towards either the major or the minor groove of the duplex. This allowed selection of the A- and B-forms of the hybrid [23].

7.2 Orientation and Dynamic Properties of DNA

7.2.1 Orientation Selection

The distances between spin labels, as well as their mutual orientation, can be determined from orientation selective measurements [24–27]. A special rigid spin label **Ç 4-20** was developed for DNA studies [28]. This label is based on cytosine, and is rigidly oriented and fixed in the DNA duplex by hydrogen bonds with cytosine's complementary base guanine:

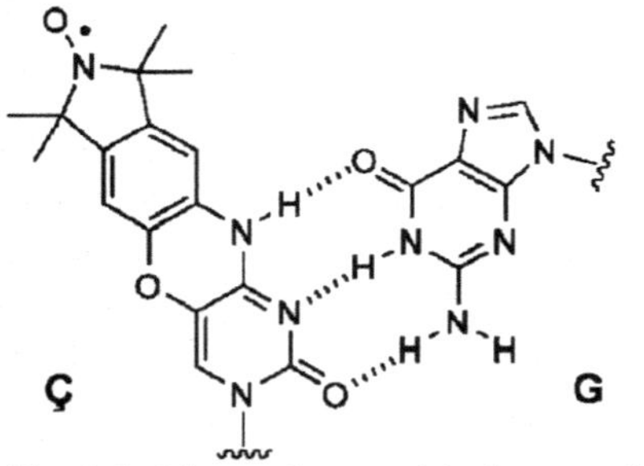

The label is coplanar with the guanine and is parallel to the other paired bases and perpendicular to the DNA helix axis, Fig. 7.9 *Left*. The angle between the vector *r* connecting the labels and the normal to the plane of each label is the same. This makes it possible to determine the angle β between *r* and the helix axis from the PELDOR *V(T)* measured with different combinations of ω_A and ω_B. A detailed description of this analysis and the experimental results are provided in [27]. The *V* (*T*) were measured at X-band for $\Delta\omega_{AB}/2\pi$ in 10 MHz steps from 40 to 90 MHz. The position of the second label was scanned from *N3* to *N14,* Fig. 7.9 *Left.* The angle β varied with the position of the label in the DNA, Fig. 7.9 *Right*. The β angle for an idealized DNA duplex corresponds well to the experimental results [27].

Several new labels are suggested for DNA investigations, Fig. 7.10, with differing rigidity when incorporated in DNA oligonucleotides [29]. $^{\mathbf{ExIm}}$**U** is the first conformaly-unambiguous spin label for nucleic acids; the nitroxyl N–O bond is coaxial with the three single bonds attaching the otherwise rigid isoindoline-based spin label to a uracil base. This label has very high rotational mobility around these single bonds in duplex DNA relative to the structurally similar spin label $^{\mathbf{Im}}$**U**, whose rotation is restricted by an intramolecular hydrogen bond. These labels were

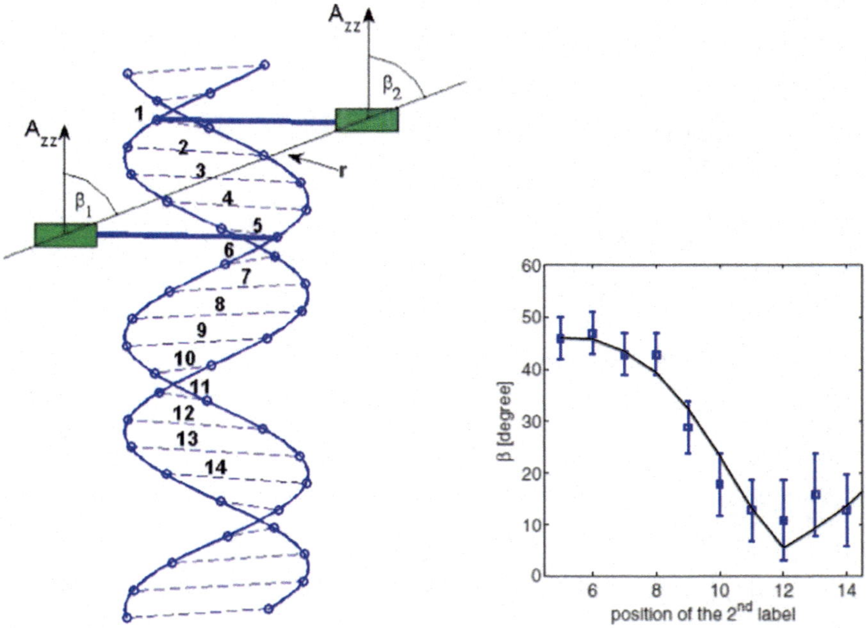

Fig. 7.9 *Left*: Orientation of spin labels in the DNA structure, spin labels are attached here to the base pairs numbered 1 and 5; the principal axis of hyperfine interaction tensor A_{zz} is normal to the label plane so the β_1 and β_2 angles coincide; *Right*: Experimental and calculated dependence of $\beta = \beta_1 = \beta_2$ on the position of second label. Reprinted from Marko et al. [26] with permission of American Physical Society, copyright 2010

used in two doubly-labeled, 14-bp duplex constructs with either seven- or ten-base pairs between the labels in **DNA(1,9)** and **DNA(1,12)**, respectively [29].

PELDOR distance measurements between pairs of **^{Im}U, ^{Ox}U**, and **^{ExIm}U** in duplex DNA showed strong orientation selection with **^{Im}U** pairs, with $V(T)$ depending on the location of ω_A and ω_B in the EPR spectrum. The orientation selection was moderate for **^{Ox}U**, and absent for **^{ExIm}U**. Thus, precise distances can be extracted from **^{ExIm}U** pairs from just a single measurement without having to take orientation selection into account. Interspin distances were obtained at X-band by Tikhonov regularization from sums of $V(T)$ with $\Delta\omega_{AB}/2\pi$ from 40 to 85 MHz [29]. The experimental distances for **DNA(1,9)** and **DNA(1,12)** constructs are: 3.43 and 3.41 nm between **^{Im}U** labels; 3.32 and 3.41 nm for **^{Ox}U**; and 4.01 and 3.58 nm for **^{ExIm}U**, respectively, in good agreement with calculated models [29].

The first comparative PELDOR study at 263 and 94 GHz examined a model 20-bp RNA duplex system with the rigid **Ç** label in positions 6 and 16 [30]. There was considerable modulation depth and significant orientation selection at 263 GHz, particularly in the $g_x - g_y$ plane of the nitroxyl. The difficulties, as well

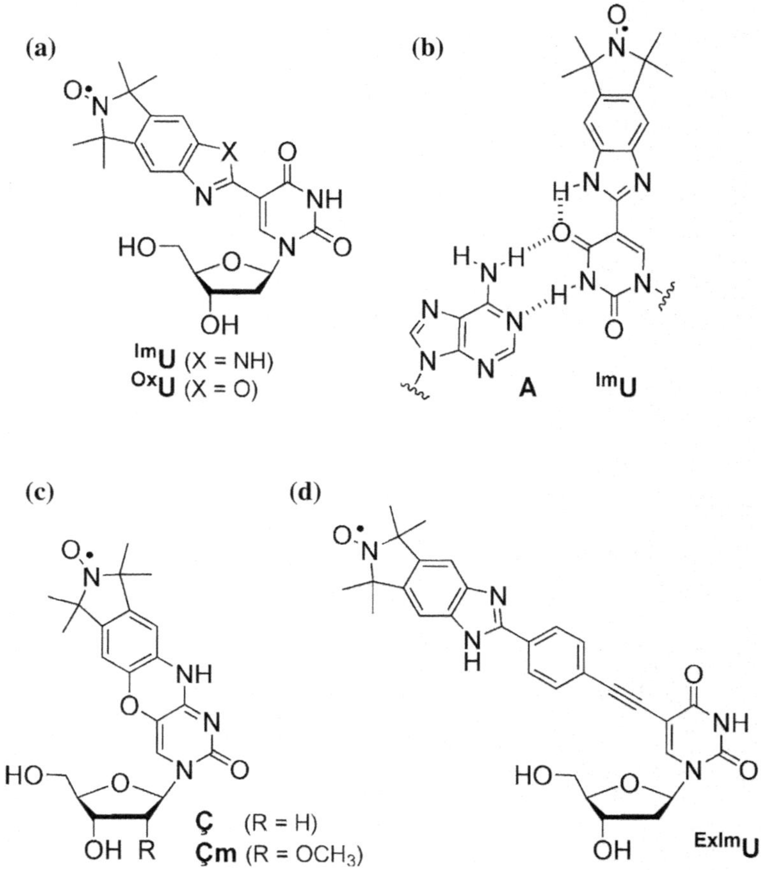

Fig. 7.10 Spin-labeled nucleosides: **a** isoindoline-derived spin labels **^{Im}U** and **^{Ox}U**, **b** base pairing of **^{Im}U** with **A**, showing the intramolecular hydrogen bond that restricts rotation around the bond connecting the nitroxyl to the base, **c** the rigid spin labels **Ç** and **Çm**, **d** the extended benzimidazole spin label **^{ExIm}U.** Reprinted from Gophane et al. [29] with permission of Wiley and Sons, copyright 2007

as the advantages, encountered at such ultrahigh frequencies were noted. There are significant, new possibilities for investigating the orientation of spin labels in structures that are more complex than simple linear single- or double-stranded nucleic acids.

7.2.2 Nucleic Acid Dynamics

The dynamic properties of nucleic acids are significant for understanding the kinetics and the mechanisms of cellular processes, e.g., replication and transcription, in which proteins twist and bend the DNA. Several urgent problems in modern biophysics center on the molecular dynamics of nucleic acids. It was believed, in the early theoretical and experimental research, that the dynamic properties of DNA duplexes could be described by the elastic cylinder model [31]. However, modern physical studies of the mobility of DNA helices [32–34] suggest at least three types of possible motions, including: (A) elongation—the helical pitch changes without any change in diameter; (B) unwinding—the diameter and length change but the pitch does not; and (C) bending—the helix bends without changes in radius or pitch.

It is repeatedly found that the $F(r)$ obtained from PELDOR measurements on rapidly-frozen glasses, corresponds quite well to the spectrum of conformations that modern MD methods predict in room-temperature liquids. This means that PELDOR can provide a snapshot of the range of dynamics in a molecular system.

These features of PELDOR were used to distinguish between the A, B and C mechanisms in a study of the conformational flexibility of doubly spin-labeled 20-bp DNA [35]. The rigid **Ç** spin label was used and paired labels were attached to 10 model duplexes with one label at one end of the duplex, and the other label was stepped one full turn down the helix. PELDOR measurements were made at X-band, $\omega_A/2\pi \sim 9$ GHz, and at G-band, ~ 180 GHz. The $V(T)$ and their dependence on $\Delta\omega_{AB}$ were measured for all the duplexes. The width $\Delta = <\Delta R^2>^{1/2}$ of the distance distribution, Fig. 7.11a, was determined by averaging the orientation selectivity data [25]. This procedure eliminates correlations between label orientations from the measured distances. In addition, orientation selection measurements were made for all spin pairs [24–26]. The $V(T)$, the linewidth of $F(r)$, and the mutual orientation of the labels were simulated for the motional model.

The relationship between the width of $F(r)$ and the position of a label was used to interpret the experimental data, Fig. 7.11b. Only the unwinding model B fit the data. The exact twisting and stretching of the helix in this model can be determined from the orientation selection data obtained in the course of the experiments at G-band. The angle between the N–O bonds of labels one turn apart varied by $\pm 22°$. As mentioned in [35], the results completely agree with the model of cooperating fluctuations, the so-called "respiratory" model of DNA duplex motion, where the pitch of a helix remains constant, but the helix radius and the length of the DNA molecule vary in a correlated manner. According to the PELDOR data, the helix radius changes by 11% while the DNA length changes by ±6%.

These PELDOR results correlate with the small angle X-ray scattering data (SAXS) [34] and with fluorescent microscopy results [32] for short DNA oligomers. The wide variety of experimental approaches used in this work [35]: the unique set of spin-labeled DNA, studies of the orientation selection, measurements at various frequency ranges; revealed the full potential of PELDOR not just for structural studies, but for thorough studies of macromolecular dynamics as well.

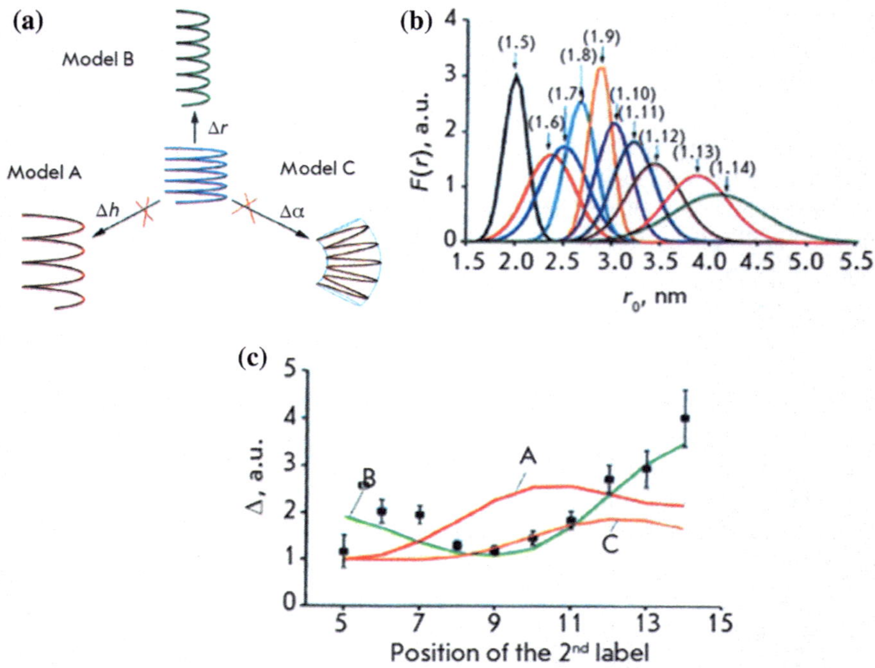

Fig. 7.11 **a** Three models of motion in duplex DNA (see text); **b** $F(r)$ from X-band PELDOR data using a Gaussian approximation for $F(r)$, with orientation selection for **Ç** spin labels pairs; **c** experimental dependence of the width, $\Delta = \langle \Delta R^2 \rangle^{1/2}$, on the distances between the labels and the theoretical simulations for mobility models A, B, and C; the minimal Δ value corresponds to the distance between 1 and 9 labels of DNA. Reprinted from Marko et al. [35] with permission of American Chemical Society, copyright 2011

7.3 Nonlinear Duplexes and Tertiary Structures

The structure of DNA and RNA is not restricted to a linear helix, but can be much more complex, similar to the tertiary structure of biopolymers. Relatively long single-stranded RNA can form duplexes between their complementary segments, while non-complementary segments form rings, hairpins, and loops, which contain several nucleotides. The distances between specific nucleotides in these secondary structures can differ dramatically from those between nucleotides in ordinary helices.

Distances between TEMPO **4-19** labels attached to the $-NH_2$ groups of specific guanine, adenine, or cytosine bases within a RNA hairpin structure 20 nucleotides in length have been measured, Fig. 7.12a [36]. The labels were positioned ten bases apart in a sequence that can fold into a hairpin structure or a duplex if a complementary strand was present. In a duplex, the labels are always 10-bp apart and PELDOR $V(T)$ give the same distance of 3.1 nm expected for A-form RNA,

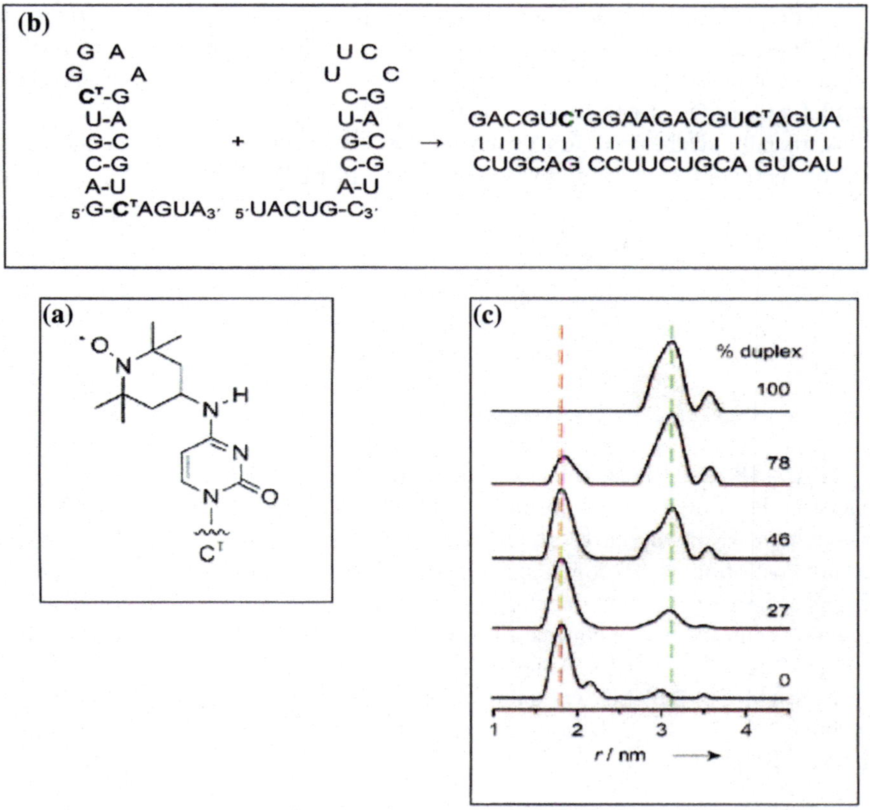

Fig. 7.12 RNA hairpin investigation: **a** structure of the spin label; **b** formation of a RNA duplex from two hairpins, the hairpin on the left is spin labeled while the other hairpin is complementary but unlabeled, when mixed, the two complementary hairpins can form a duplex; **c** PELDOR distance spectra in the mixtures as the unlabeled, complementary hairpin is titrated into a solution of the labeled hairpin, the distance spectra show the change in relative amount of the RNA duplex Reproduced from Sicoli et al. [36] with permission of John Wiley and Sons, copyright 2010

regardless of the labeled nucleotide type or the position of the label pair within the RNA duplex.

In this RNA hairpin structure of 20 nucleotides, six complementary nucleotides form a double helix as the hairpin stem, four nucleotides form the loop, and the remaining nucleotides are in a single strand. A duplex with labels only in one strand was formed after a completely complementary RNA molecule without spin labels had been added to the system, Fig. 7.12b. Samples containing different amounts of the labeled RNA and its unlabeled complementary RNA were measured by PELDOR. The distance spectrum obtained in these experiments, Fig. 7.12c, indicates that in frozen buffer solutions, spin-labeled hairpins with 1.8 nm between labels are present, along with duplexes having 3.1 nm between labels.

Single RNA strands can form more complex structures, such as rings, or semi-rings, in conjunction with hairpins. These structural elements may respond to certain molecules; they are known as RNA riboswitches and play a crucial role in regulating transfer of genetic information in cells.

Artificially-synthesized RNA riboswitches consisting of 27 nucleotides, that accept neomycin, have been investigated using PELDOR [37]:

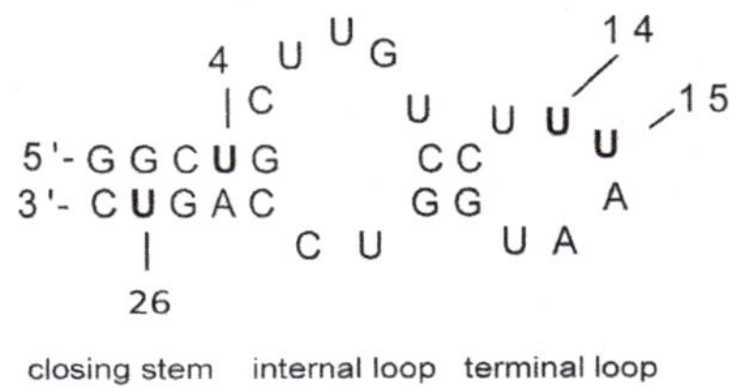

TPA **4-18** spin labels were introduced into the bolded uracil residues and the interspin distances between positions 4–14, 4–15, 14–26, 15–26 were determined. An attempt was made to determine the conformational changes in this riboswitch during formation of a neomycin complex [37]. It was found, however, that the distance between the spin-labeled uridine bases were virtually the same in this complex as in the neomycin-free riboswitch. This result suggests that the structure has good conformational stability [37].

The rigid **Ç 4-20** label was an ideal probe for PELDOR studies on the cocaine aptamer, which is a DNA three-way junction that binds cocaine at its helical junction [38]:

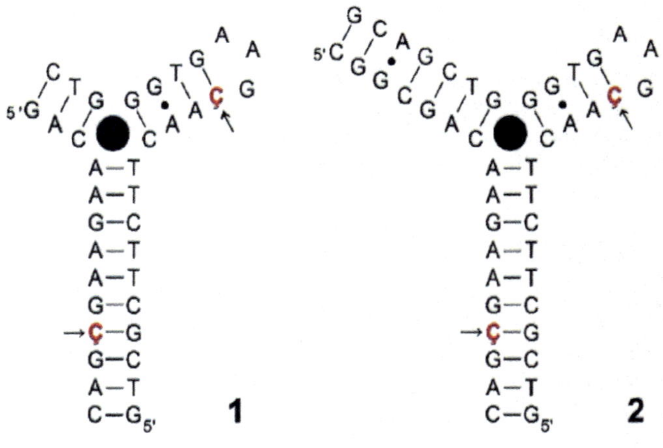

Here two aptamers were studied. Both have the same sequence and are labeled at the same positions, the difference is a longer side helix for aptamer **2**. The filled black circle represents the cocaine ligand. The positions labeled with **Ç** are marked by arrows.

The conformation and overall flexibility of the aptamer in the presence and absence of cocaine was studied at X- and Q-bands [38]. The **Ç** label was incorporated pairwise into two helices of the aptamer. PELDOR orientation selection experiments determined the mutual orientation and the distances between pairs of labels. The cocaine-bound and non-bound states differ in their conformational flexibility, which is lower with cocaine bound. There were only small changes in the width and mean value $\Delta r = 0.3$ nm of $F(r)$ upon cocaine binding. A detailed simulation of $V(T)$ revealed considerable conformational flexibility in the helical junction, even after binding to cocaine.

Mg^{2+} ions had a major effect on the structure of the "hammerhead ribozyme," HHRz [39]. The distance spectrum for TEMPO **4-19** spin labels incorporated in two stems of the HHRz structure showed that the addition of Mg^{2+} ions stabilized a tightly-folded structure with the labeled stems quite close ~2.4 nm. Without Mg^{2+} ions or with mutations in other parts of the ribozyme, no ordered structure was observed. It is noteworthy that Mg^{2+} ions are required for the catalytic activity of this HHRz.

Changes in distances between spin labels that are due to conformational transformations were also seen by PELDOR in other, more complicated RNA and DNA molecules having a range of tertiary structures, see, e.g., [40, 41]. This important area in structural biophysics is ripe for innovative PELDOR studies.

7.4 PELDOR Inside Cells

In-cell studies are vital to determine the effect of the intracellular environment and conditions, e.g., viscosity, molecular crowding, the activity of water, interactions with other macromolecules, etc., on the structure of nucleic acids. Such information could be found by PELDOR distances measurements on macromolecules or complexes that contain more than one intrinsic spin or on doubly spin-labeled nucleic acids or proteins introduced into cells.

Sample preparation for EPR studies of doubly spin-labeled nucleic acids or proteins introduced into whole cells is a rather complicated procedure [42–45]. The major approach is the injection method where a large specialized cell, having a diameter about 1 mm, is injected with a solution containing the spin-labeled macromolecule without destroying the cell. This needs to be repeated dozens of times. There is special equipment for microinjection: a glass micropipette with the needle of 0.5–5 μm mounted on a pneumatic or oil-driven injection device is used under a microscope. A detailed description of sample preparation by microinjection can be found in [46]. Although this approach allows the study of a much wider range of nucleic acids and proteins that does the intrinsic radical approach, there can be questions concerning how well the cells, typically oocytes, represent normal physiology of typical cells; how well the injected macromolecules are transported into the proper intracellular compartments; and how much the injected material perturbs the physiological state of the cell.

The main obstacle in conducting in-cell PELDOR experiments is the short half-life of nitroxyl spin labels in the intracellular environment. They are readily reduced by reducing agents, such as small molecules or enzymes, which decrease the EPR signal and convert doubly-labeled molecules into mono-labeled molecules.

A detailed investigation of the decay kinetics of spin labels in cells is given in [47]. The five-member ring, similar to SAT **4-15**, and the six-member ring TOAC **4-22** nitroxyl radicals were studied in *Xenopus leavis* oocyte extracts. The nitroxyl radical stability primarily depended on the size of the heterocyclic ring. The half-life for the five-membered ring nitroxyl is 150 min but for TOAC is 0.7 min. A kinetic analysis of the nitroxyl reduction in cell revealed an enzyme-mediated reaction [47]. These findings in oocyte extracts are important for the selection of the nitroxyl label for in-cell experiments, but results may be different for other cell types, nitroxyl derivatives and conditions. TPA **4-18** is reduced in oocyte extracts about four-fold faster than when it is attached to RNA or DNA [48].

Cells need to be incubated about 1 h at physiological temperatures after microinjection [49], making signal-to-noise an important issue for any in-cell PELDOR measurements. The incubated cells are usually shock-frozen in liquid nitrogen prior to PELDOR experiments made at cryogenic temperatures, but loss of doubly-labeled molecules during incubation is a major concern. A detailed protocol for PELDOR measurements in cells can be found in [46].

7.4.1 Ubiquitin

The first in-cell PELDOR experiment was reported in 2010 [49]. Two doubly-labeled constructs of human ubiquitin, a small regulatory protein, were made with five-membered-ring spin labels and injected into *X. laevis* oocytes. PELDOR from the two constructs found distances of 3.14 and 2.60 nm in the oocytes, while the in vitro values were 3.11 and 2.65 nm, respectively. The width of the distribution function $F(r)$ increased in the oocytes after a 1 h incubation at 18 °C. This broadening was attributed to some intracellular events but not to changes in the ubiquitin conformation [49]. Ubiquitin has also been studied using Gd(III) labels, see Sect. 7.4.4.

7.4.2 Nucleic Acids

PELDOR measurements were made for 12-bp double-labeled DNA in vitro and in cells [48]. The PELDOR $V(T)$ of in-cell DNA have a somewhat different pattern of oscillation than in vitro. The $V(T)$ in vitro can be fitted by $F(r)$ with a single Gaussian peak at 2.1 nm and a width of 0.14 nm, while the in-cell $V(T)$ requires two Gaussian peaks in their $F(r)$. The shorter distance matches the in vitro measurement, indicating that the duplex structure is undisturbed in cells. The longer distance of 3.7 nm has a very broad width of 2 nm and may arise from the stacking of DNA molecules in cells. End-to-end stacking of 6- to 20-bp nucleic acid duplexes and ordering into semirigid rod-shaped structures is well known [50, 51].

Although in-cell reduction of nitroxyl labels attached to DNA occurs at the same rate as when attached to RNAs, the PELDOR modulation depth for DNA was deeper. This increase in modulation depth was ascribed to dipolar interactions of more than two coupled spins, which could occur with end-to-end stacking of the duplexes. This explanation is further supported by the background V_{INTER} which shows a one-dimensional spin distribution with $d = 1$ Eq. 1.19 as expected for linear stacking of the DNA [48].

The 14-bp cUUCGg tetraloop hairpin RNA and the 27-bp neomycin-sensing riboswitch were then studied in *X. laevis* oocytes and compared with in vitro PELDOR measurements [48] and with structures of the unlabeled RNA [52, 53].

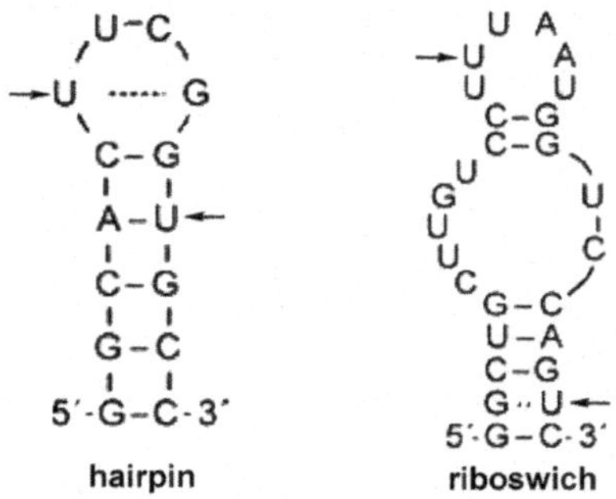

TPA **4-18** spin labeling at positions shown by arrows took about 50 min. In vitro PELDOR experiments were carried out in intraoocyte buffer. Distances of 1.8 nm for the tetraloop RNA hairpin and 3.4 nm for the neomycin riboswitch were obtained from the in vitro *V(T)*. The widths of the *F*(*r*) were 0.27 nm for the hairpin and 0.7 nm for the riboswitch. Practically identical PELDOR results were obtained for both RNAs after 10 and 70 min of incubation. The fact that the distances measured in the in vitro and in-cell experiments were the same implies that the 14-mer hairpin RNA and the 27-mer riboswitch have stable global structures.

7.4.3 G-Quadruplexes

DNA sequences with a high degree of polymorphism are of particular interest for in cell PELDOR measurements because PELDOR can identify different conformations of the DNA constructs. One such important structure is the human telomeric G-quadruplex. The G-quadruplexes are tertiary structures formed by nucleic acids with sequences rich in guanine. Four guanine bases can associate to form a square planar structure, a tetrad, and two or more of these tetrads can stack on top of each other to form a G-quadruplex. The structure is stabilized by a cation, especially K^+,

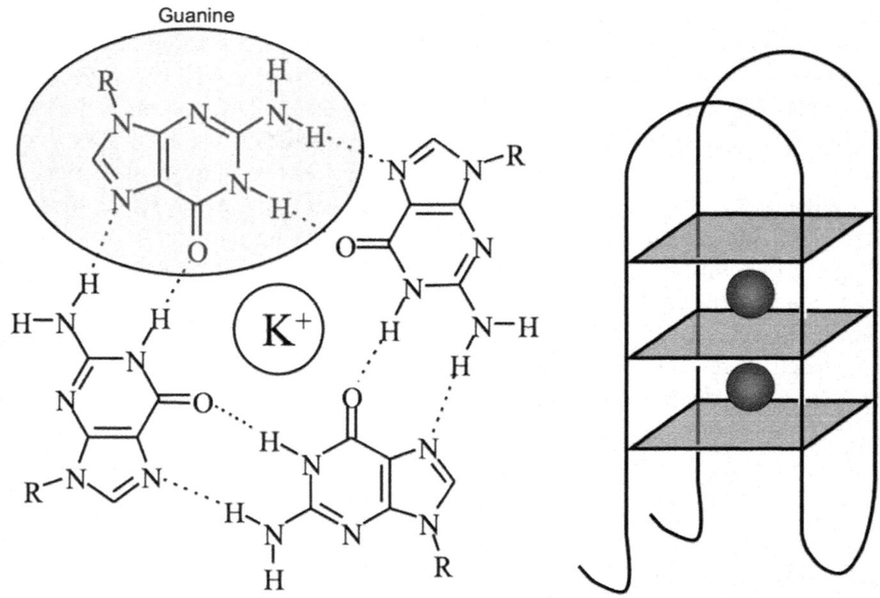

Fig. 7.13 Schematic outline of the structure of a G-quadruplex *left*: a G-tetrad; *right*: intramolecular loops of nucleic acids linking G-tetrads are shown

sitting in the central channel between each pair of tetrads, Fig. 7.13 [54]. The quadruplex, in response to subtle changes in environmental conditions, is known to adopt drastically different conformations. In particular, its conformation and the mixture of conformations depends upon its counterions, K^+ versus Na^+ and on the DNA sequence around the quadruplex.

The human telomeric quadruplex was studied in buffer with K^+ ions [55]. The conformation in this case is known from NMR and X-ray analysis, Fig. 7.14 [56–58]. In the PELDOR experiments, the DNA sequence d[A(GGGTTA)$_3$GGG] was labeled by TEMPA **4-17** in positions 5 and 11, as indicated by a bold **T**:

5'—A-G-G-G-**T**-T-A-G-G-G-**T**-T-A-G-G-G-T-T-A-G-G-G—3'

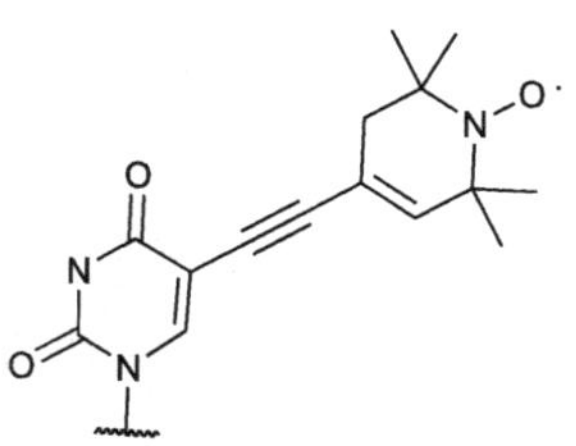

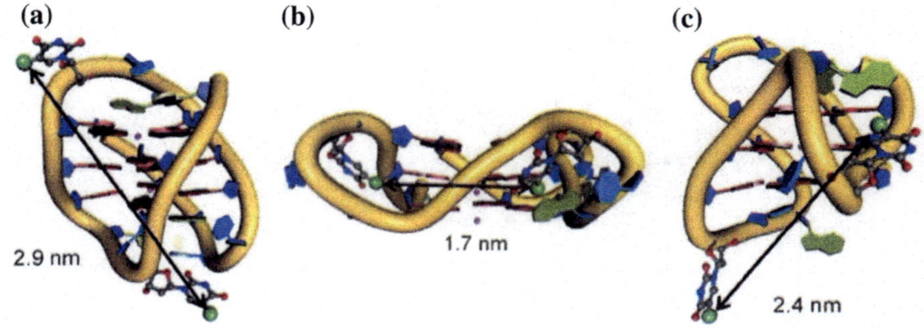

Fig. 7.14 G-quadruplex topologies based on X-ray and NMR structures: **a** the antiparallel basket quadruplex in the presence of Na^+ ions [56], **b** the parallel propeller form in the presence of K^+ ions [57], **c** NMR structure of a hybrid 3 + 1 quadruplex of a slightly modified sequence in K^+ ions [58]

Table 7.3 Structural data for G-quadruplex conformations [55, 59]

Conditions	PELDOR experiment (nm)		PDB based calculations (nm)	
	Propeller	Basket	Propeller	Basket
Buffer K^+	1,8 ± 0.2 $\Delta \leq 1.0$	3,0 ± 0.1 $\Delta \leq 1.0$	1.7	2.9
Cells	2.0 ± 0.1 $\Delta = 0.7$	2.9 ± 0.1 $\Delta = 0.7$		

Tikhonov regularization analysis of the PELDOR *V*(*T*) gave two broad lines at the distances corresponding to the expected value for the parallel propeller form, whereas the second species has a distance of identical to the antiparallel basket quadruplex, Fig. 7.14a, b and Table 7.3. These conformers are present in a 1:1 ratio in the frozen solution.

The same human telomeric sequence was synthesized and spin-labeled with the more stable TPA **4-18** label [59]. It was injected into the animal hemisphere cytoplasm of stage VI oocytes of *X. laevis* where nucleus-like conditions prevail [60]. For the distance measurements, the telomeric sequence again was labeled in the trinucleotide loop region at the positions 5 and 11 [55, 59]. And again, two distances were found from analysis of *V*(*T*) using a Gaussian line shape, Table 7.3. One distance of 2.9 nm corresponds to the antiparallel basket quadruplex, the second distance of 2.0 nm to the parallel propeller form.

Evidence for a 3 + 1 conformation of a G-quadruplex was obtained in later work [61]. In contrast to previous studies [55, 59], the spin labeled quadruplex was embedded in a long stretch of human telomeric DNA $d[A(G_3T_2A)_{11}G_3]$ containing three quadruplex units. The DNA was labeled by TEMPO **4-19** attached at positions 29 and 35 of the DNA sequence. In the presence of Na^+ ions, the distance between the two nitroxyl spin labels in the middle quadruplex unit of this long DNA was $r = 3.0 \pm 0.1$ nm which corresponds to the antiparallel basket

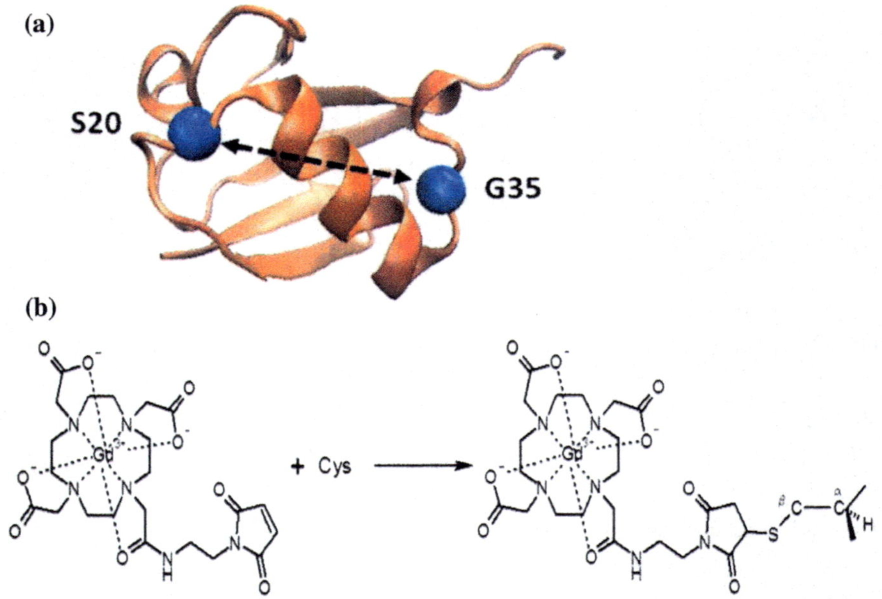

Fig. 7.15 **a** Ribbon structure of the ubiquitin (PDB 1UBQ), Ser20 and Gly35 were substituted with cysteines and labeled with Gd(III)-DOTA-Maleimide; **b** the labeling reaction. Reprinted from Martorana et al. [62] with permission of American Physical Society, copyrights 2014

quadruplex topology. However in K^+ solution, $r = 2.5 \pm 0.1$ nm, corresponding to the (3 + 1) hybrid topology, Fig. 7.14c. These results show that PELDOR should prove very useful in further investigations of quadruplex structures and interactions in complex mixtures and even complex cellular systems.

7.4.4 Ubiquitin and Gd(III)

An innovative approach for PELDOR in-cell investigations features: (1) novel Gd (III)-based spin labels and protein linkers that are stable under in-cell conditions; (2) high-field measurements that feature high sensitivity with Gd(III), thus allowing the detection of low concentrations and use of small sample sizes; (3) efficient delivery of the labeled protein into cells by hypo-osmotic shock, reaching in-cell concentrations sufficient for PELDOR measurements while maintaining cell viability; and (4) D_2O exchange to slow echo decays [62, 63].

These advantages were demonstrated on doubly-labeled ubiquitin in HeLa cells. A mutant of human ubiquitin with two cysteines replacing Ser20 and Gly35, Fig. 7.15a, was doubly labeled with Gd(III)-DOTA maleimide **4-25** [64], Fig. 7.15b, and introduced into HeLa cells by osmotic shock [65, 66]. Operational simplicity, time efficiency, and high probability of cellular viability are attractive advantages of this technique [62]. Partial deuteration of the intracellular medium by <20% D_2O was used to slow phase relaxation.

The 4pPELDOR data and analysis are shown in Fig. 7.16. The distance distribution obtained from both in vitro and in-cell samples is similar: it has a maximum around 3.2 nm and a width of about 1.5 nm. This interspin distance distribution is in good agreement with previous in vitro X-band PELDOR measurements on S20C/G35C labeled with conventional nitroxyl spin labels [49, 67]. No comments were made about the right-hand shoulder in $F(r)$ in-cells.

Significant differences were observed in the background $V(t)$, which is substantially stronger in-cell than in buffer and indicates a higher local concentration in-cell of 110 μM. This value exceeds considerably the bulk concentration of 4.5 μM and is larger than that used with hypo-osmotic shock, 100 μM, suggesting localization and inhomogeneous distribution in the cell, in agreement with the fluorescence measurements [62]. The physiological relevance of such localized proteins needs to be carefully considered.

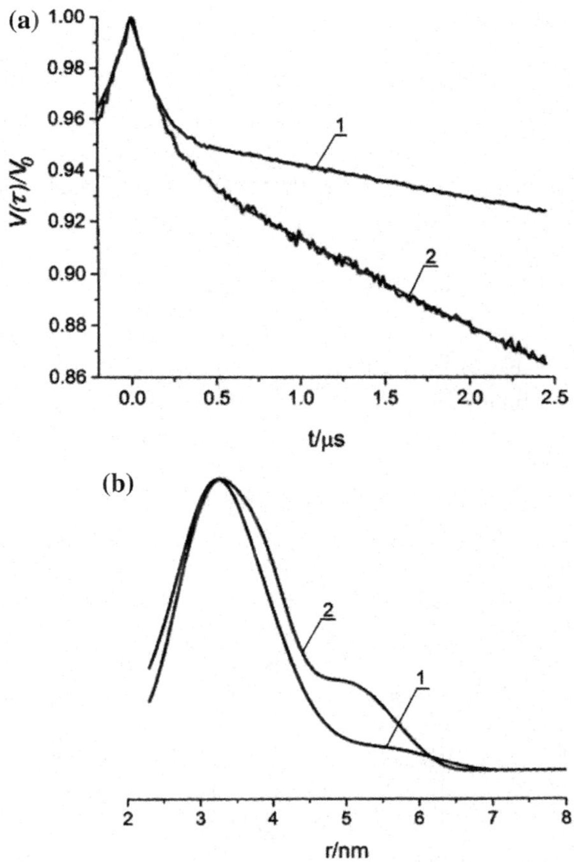

Fig. 7.16 W-band PELDOR results of doubly-labeled ubiquitin *in vitro* and in-cell: **a** experimental DEER trace and background fit: *1* in vitro at 25 μM, and *2* in-cell; **b** distance distribution $F(r)$ obtained using Tikhonov regularization for *1* in vitro, and *2* in-cell. Reprinted from Martorana et al. [62] with permission of American Physical Society, copyright 2014

Nevertheless, studies to date in cells and complex systems show that measurements of biomolecule structure and organization by PELDOR has a promising future in biology.

7.5 DNA Lesions

The effect of various lesions on DNA structure, in particular, the changes in the distances between spin labels, was studied by PELDOR [68–70]. Lesions in DNA are caused by many factors, such as chemical modification of the DNA at a certain location, nucleotide substitution, nicks in one of the DNA strands in the duplex, etc. All these lesions that chemically modify DNA can cause mutations or cell death. The physical consequences of these lesion on structure and flexibility of the DNA are of particular relevance for understanding the biological effects of these lesions and how they are recognized and repaired.

7.5.1 Base Mismatch

Duplexes containing AA and TT mismatches, i.e., non-canonical pairs, were studied by attaching two TEMPO **4-19** spin labels onto nucleotides on each side of the mismatch [71]:

```
5'—T-A-G-G-T-C A* X-Y A*·C-T—3'
3'—A-T-C-C-A-G-T-Z-W-T-G-A—5'
```

where **A*** is 7-deazaadenozine containing the TEMPO spin label at C7; X-Y is the noncanonical pair dA-dA or dT-dT at positions 8 or 9 of the duplex. The distance between the spins in the canonical duplex, when XY/ZW = AT/TA, was $r = 1.83$ nm. The distance for the duplexes TT/TA and AT/AA containing the TT or AA noncanonical pair at position 8 was $r = 1.73$ nm. However, $r =$ 1.87 and 2.08 nm if the duplexes contained noncanonical pairs (AT/TT and AA/TA, respectively) at position 9. Thus, the introduction of a noncanonical pair into position 8 reduces the interspin distance, while the introduction into position 9 increases it as compared to the canonical duplex. The DNA mismatch alters the structure of the adjacent base pairs, but in different way: causing compression or expansion in the two cases examined here.

DNA duplexes containing three TEMPO spin labels were also studied [72]. These **Y** labels were attached to the *C5* atom of uridine residues of an alkynyl-oligonucleotide using "click chemistry". A tetrahydrofuran lesion F was introduced into one of the DNA strands in addition to the spin labels. So, each DNA duplex contained three spin labels and one damaged site.

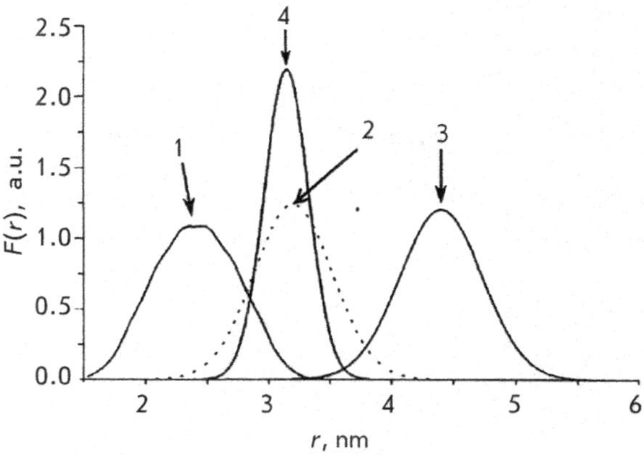

Fig. 7.17 *F*(*r*) for a triply-labeled DNA duplex (*1*, *2*, *3*) and after treatment with Endo IV (*4*), changes are attributed to double-labeled DNA. Reprinted from Flaender et al. [72] with permission of Wiley and Sons, copyright 2007

5'—G-Y-A-A-G-G-T-C-C-A-G-F-G-A-Y-G—3'
3'—C-A-T-T-C-C-A-G-G-Y-C-T-C-T-A-C—5'

The PELDOR *V*(*T*) in this 3-spin system were analyzed using the conventional procedure [73], modified for a 3-spin system [74]. As expected, *F*(*r*) for this system consisted of three lines with peaks at *r* = 2.50, 3.15, and 4.60 nm and the widths of 0.05, 0.45, and 0.75 nm, respectively, Fig. 7.17.

The interaction between this DNA duplex and apurinic/apyrimidinic AP endonuclease IV (*EndoIV*) from *Escherichia coli* was also investigated [72]. *EndoIV* cleaves the damaged strand [75] at the AP site and at AP site analogs containing F residues from damaged deoxyribose. The duplex spin-labeled DNA was cut at the F residue, yielding a duplex containing only two spin labels and a short fragment with one label and the F

5'—F-G-A-Y-G—3'

5'—G-Y-A-A-G-G-T-C-C-A-G—3'
3'—C-A-T-T-C-C-A-G-G-Y-C-T-C-T-A-C—5'

A single line was detected in *F*(*r*) after treatment with *EndoIV*, Fig. 7.17, with *r* = 3.20 nm and a width of 0.75 nm, which was attributed to the spin-labeled duplex that remained after cleavage of the damaged strand. The results obtained demonstrate the potential for using PELDOR in the investigation of 3-spin DNA systems and, more importantly, expand the range of systems that can possibly be used to probe the interaction of DNA with proteins and enzymes.

Changes in the inter-label distances in DNA duplexes containing other structural defects and insertions in one of the duplex strands has also been studied [68]. The 20-bp DNA

Table 7.4 Structures of radical R, nucleotides, and non-nucleotide insertions [69, 70]

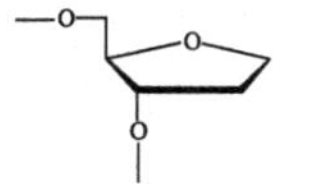	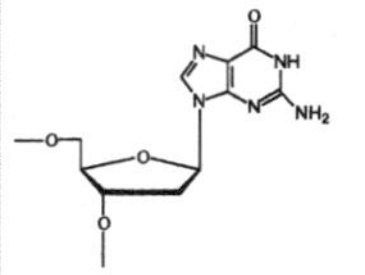	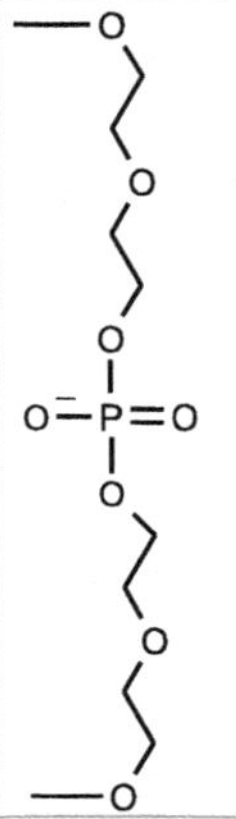	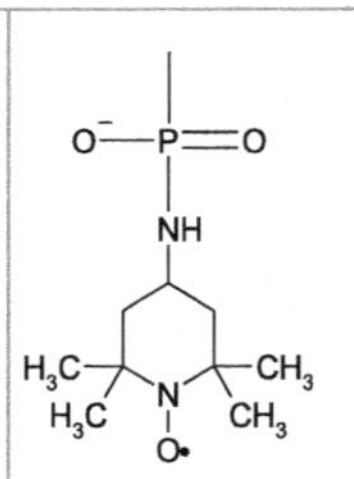	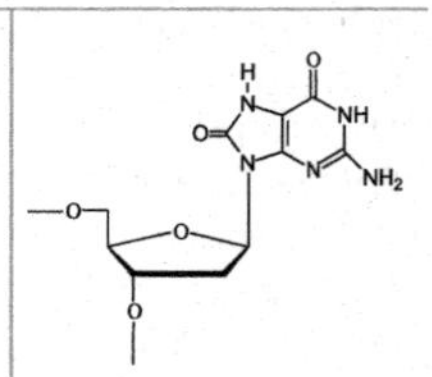
THF (F)	**G**	**deg2p**	**R**	**8oxoG**

5'—G-A-C-C-C-T-G-C-A-T-C-T-A-G-G-A-T-G-C-C—3'
3'—C-T-G-**G**-G-A-C-G-T-A-**G**-A-T-C-C-T-A-C-G-G—5' (A)

5'—G-A-C-C-C-T-G-C-A-T-C-T-A-G-G-A-T-G-C-C—3'
3'—C-T-G-**G**-G-A-C-G-T-A-G-A-T-C-C-T-A-C-**G**-G—5' (B)

with TEMPO spin labels attached to guanine residues of one strand at positions (4;11) (**A**) and (4;19) (**B**); the second strand of the duplex contained various lesions —nicks, gaps, modified nucleotides, and bulges, Table 7.4. Two of the non-nucleotide insertions (**Pr, Et**) are:

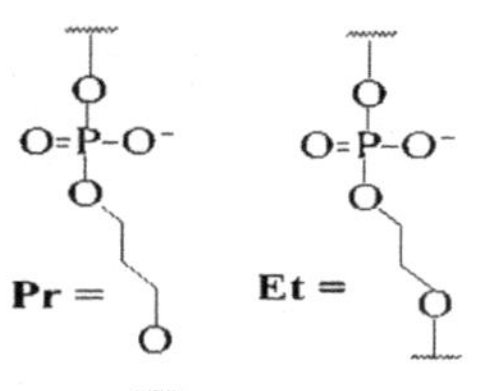

4pPELDOR at 60 or 70 K was used with conventional data analysis methods in each case. The samples contained 50 or 100 μM of spin-labeled DNA in saline with 15–20% glycerol. The *F*(r) for the duplex **A** (4;11), undamaged and with various lesions, show a variety of changes, Fig. 7.18. In order to eliminate the orientation selectivity, the PELDOR *V*(*T*) was averaged over 10 measurements by variation of the magnetic field across the EPR spectrum. These measures allow one to reliably assess the error in determining the average distance and the width of lines in the distance spectrum with an error ~10% [68].

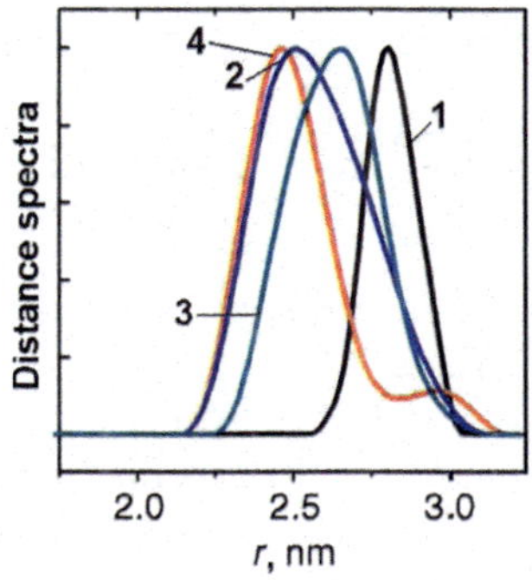

Fig. 7.18 F(r) for undamaged DNA **A** *1* with labels in positions 4 and 11, and the changes when lesions *2* **Pr**, *3* **Et**, and *4* **F** are introduced in the DNA. Reproduced from Sicoli et al. [68] with permission of Oxford University Press, copyrights 2009

The results obtained for the duplexes labeled at positions (4;11) fall into two groups [68]. The first group contained structural lesions **F**, **Pr**, and **Et** at position 7 of the second strand. A significant decrease, from $r = 2.81$ nm in undamaged **A** to 2.48 nm for **Pr**, along with a broadening of the F(r) peak from 0.21 to 0.46 nm with the appearance of some asymmetry, was found in the this group. A similar situation was observed for duplex **B** labeled at positions (4;19), where the distances changed from $r = 5.21 \pm 0.04$ nm for the undamaged structure to 5.02 ± 0.03 nm for the damaged duplexes. The width of the distribution function for duplex **B** changed from 0.33 ± 0.02 to 0.44 ± 0.05 nm. In the second group of lesions, **8oxoG**, nick, gap, and bulge, the distance changes were insignificant and due mostly to measurement errors.

When discussing the results, it is usually assumed that the width of the $F(r)$ peak characterizes the conformational flexibility of the duplex while changes in peak position reflect conformational changes. MD calculations [68] also concluded that the significant changes in r in the first group of damaged duplexes could be attributed to local changes in the conformations at lesions sites and in the complementary nucleotide of the duplex.

In general, the changes in the damaged DNA in a number of cases [68] are in qualitative agreement with the data obtained by methods such as NMR. The combination of pulse EPR spectroscopy with MD techniques for spin labeled DNA is complementary to conventional methods, such as NMR, CD, FRET, and X-ray crystallography, and provides additional information method about DNA lesions and the weak interactions of DNA with other molecules and complexes.

The sensitivity of PELDOR parameters to changes in nucleic acid structure became considerably greater when the spin labels were attached at the termini of relatively short oligonucleotides and their duplexes [69, 70]. Synthetic 12-bp oligonucleotides and their duplexes contained TEMPO **4-19** labels on their 5′- and 3′-terminal phosphate groups:

5'—R-C-T-C-T-C-G-C-C-T-T-C-C-R—3'(sG/C12)

5'—R-C-T-C-T-C-G-C-C-T-T-C-C-R—3'
3'—G-A-G-A-G-C-G-G-A-A-G-G—5' (G/C12)

The G nucleotide located in the center of the G duplex was modified by various insertions and substitutions, Table 7.4.

The spin-labeled DNA was studied in frozen, glassy water/glycerol solutions at 77 K. The *F(r)* were determined from the experimental background-free *V(T)* using Tikhonov regularization with the use of both the standard algorithm and the Gaussian approximation. The *F(r)*, Fig. 7.19a and Table 7.5, demonstrate a 2- to 3-fold narrowing for duplex DNA compared to single-stranded DNA. The insertion of nucleotide analogues results in a reduction in the average interspin distance in the duplexes. In case of the **deg2p** insertion, a noticeable broadening in *F(r)* as compared to **G/C12** duplexes was observed. It is obvious that the sharpening in the *F(r)* can be attributed to the formation of the DNA double helix, which is characterized by a more rigid structure than single-stranded DNA. The observed width of *F(r)* for the undistorted duplex **G/C12** was apparently caused by the random orientation of spin labels due to rotation around the P–N bonds. Considering the fact that the distance between the nitrogen atom and the N–O moiety of the spin label is ~0.4 nm, the maximum broadening from the reorientation of spin labels would be 1.6 nm, while the experimental width of $\Delta = 0.98 \pm 0.1$ nm for the undamaged duplex lies in this range. Effects associated with orientation selection in PELDOR

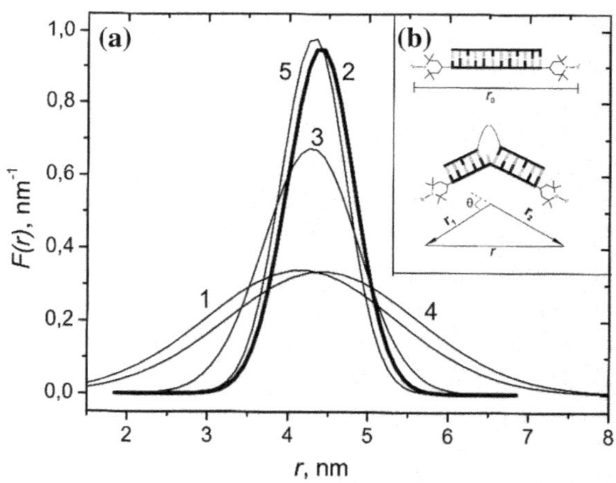

Fig. 7.19 **a** Gaussian approximation of *F*(*r*) for two spin labels in DNAs: *1* single-stranded oligonucleotide ssG, *2* thick line, duplex dsG, *3*: duplex dsG-looped, *4* single-stranded oligonucleotide ssF, *5* duplex dsTHF; **b** schematic representation of spin-labeled DNA *top*: without, and *bottom* with non-nucleotide insert, demonstrating bending and shortening of the distance between labels. Reproduced from Kuznetsov et al. [69] with permission of Royal Society of Chemistry, copyrights 2009

Table 7.5 Parameters of the distance spectra (nm) for 12-bp oligonucleotides and their DNA duplexes [69]

Sample[a]	Average distance, r, nm	Width Δ, nm
1. sG/C12	4.05 ± 0.05	2.8 ± 0.2
2. s**F**/C12	4.32 ± 0.05	2.85 ± 0.2
3. G/C12	4.35 ± 0.03	0.98 ± 0.1
4. deg_2p/C12	4.23 ± 0.03	1.39 ± 0.1
5. THF/C12	4.26 ± 0.03	0.95 ± 0.1

[a]The symbol **s** denotes single-stranded DNA

were not observed [68–70], which can also presumably be attributed to the broad, uncorrelated range of spin label orientations.

The introduction of non-nucleotide insertions into the duplex structure affects the average interspin distance. A decrease in the distance raises the possibility of bending of the duplex at the insertion site, due to the additional degree of freedom at the insertion site, Fig. 7.19b. The experimental r_0 in the intact duplex and r for the distorted duplex can be used to estimate bends with θ = 23° in the **F/C12** duplex and θ = 27° in the looped **deg2p/C12** duplex.

Any increase in the length and flexibility caused by the insertion in **deg2p/C12** adds additional broadening in the $F(r)$: Δ = 1.39 nm. The broadening is too large to simply attribute it to the spread in the bending angle of the duplex without taking into account elongation of the duplex. Hence, the bending angle estimated for this insertion is only a lower bound.

Incorporation of a non-nucleotide lesion in a DNA duplex opens the possibility of duplex bending. An increase in the number of bonds in the lesion increases its flexibility. The increase in duplex flexibility near the insertion causes a considerable increase in the range of bending angles and total duplex lengths.

The same approach was used to study lesions in even longer 13-bp and 17-bp DNA duplexes [70]. However, the main aim of the work was to investigate the conformational transformations of DNA during its interaction with the formamidopyrimidine-DNA-glycosylase or Fpg protein from *E. coli*. Fpg is considered to be one of the key factors in the process of DNA repair. The investigated duplexes are:

```
3'—R-A-G-A-G-A-G-C-G-G-A-A-G-G-R—5'
   5'—T-C-T-C-T-C-G-C-C-T-T-C-C—3'  (G/C13)

3'—R-A-G-A-G-A-G-A-G-C-G-G-A-A-G-G-A-A-R—5'
   5'—T-C-T-C-T-C-T-C-G-C-C-T-T-C-C-T-T—3'
                                        (G/C17)
```

The structures of the R spin labels, nucleotides, and non-nucleotide insertions are presented in Table 7.4.

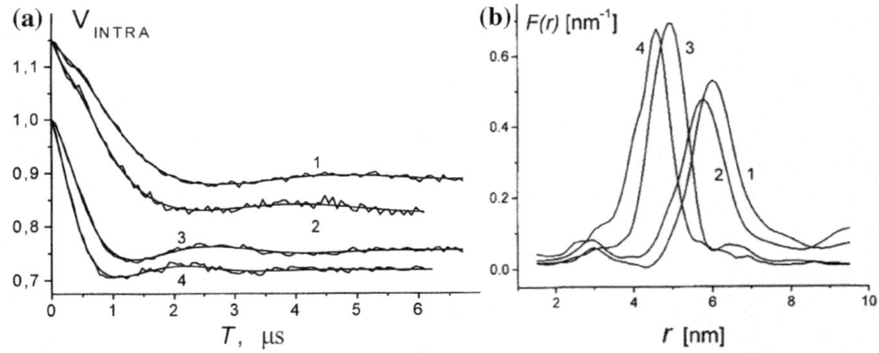

Fig. 7.20 The $V_{INTRA}(T)$ for adducted DNA and Fpg. **a** *1* and *2* refer to DNA G/C^{17} and THF/C^{17}/Fpg, respectively, *3* and *4* refer to DNA G/C^{13} and THF/C^{13}/Fpg, respectively, *1* and *2* are shifted upwards 0.15, the smooth curves are calculated using the $F(r)$ **b** the $F(r)$ between labels neglecting orientation selection, *1* and *2* refer to DNA G/C^{17} and THF/C^{17}/Fpg, respectively, *3* and *4* refer to DNA G/C^{13} and THF/C^{13}/Fpg, respectively. Reproduced from Kuznetsov et al. [70] with permission of Royal Society of Chemistry, copyright 2011

Table 7.6 Parameters (nm) of the distance distribution function $F(r)$ between two spin labels in the DNA duplexes [70]

Sample	r_{max}, nm[a]	Δ, nm[a]
G/C13	4.96	1.12
8-oxoG/C13	4.96	1.15
THF/C13	4.83	1.12
THF/C13/Fpg	4.60	1.2
G/C13/Fpg	4.78	1.1
G/C17	6.00	1.2
8-oxoG/C17	6.02	1.25
THF/C17	5.98	1.23
THF/C17/Fpg	5.76	1.2
G/C17/Fpg	5.99	1.4

[a] r_{max} and Δ were measured with errors of 0.8 and 10%, respectively

The analysis of the $V(T)$, Fig. 7.20a, gave the distance distribution $F(r)$, Fig. 7.20b, the distance at the peak maximum, r_{max} with 0.8% accuracy, and the width of the line at half-height Δ with 10% accuracy, Table 7.6.

The positions of the spin labels at the 5′- and 3′-terminal ends of the complementary second oligonucleotide of the DNA duplex made them sensitive to DNA curvature caused by either the damaged sites or the formation of complexes with an enzyme. The Fpg protein from *E. coli* caused bending even in the undamaged 13-bp duplex. A similar result has been obtained from the X-ray structure of Fpg from *Bacillus stearothermophilus* [70, 76, 77]. However, no bend was detected for the undamaged 17-bp duplex in the presence of Fpg. This could be attributed to the fact that the enzyme occupies a 10-bp DNA segment and cannot move or slide along the short DNA duplex, while sliding is possible during binding to the 17-bp DNA duplex, so Fpg may bind differently to the two undamaged duplexes. It cannot be ruled out, however, that the conformational mobility of the spin labels on the 17-bp duplex is greater than on the 13-bp duplex, which is supported by an increase in the width of $F(r)$ for the 17-bp duplex, Table 7.6.

In free DNA duplexes containing **8oxoG**, changes in the interspin distances relative to the undamaged duplex have not been seen. This is not surprising, since **8oxoG** hardly changes the DNA structure. Duplexes containing the major cyclic **F** lesion show a considerable reduction in the interspin distance, a result supported by MD computer simulations [70].

During the interaction between the duplexes and Fpg from *E. coli*, bending occurred in both the 13-bp and 17-bp duplexes. This result correlates with the X-ray structures for the cross-linked adduct of Fpg and the AP site [76–78]. It is important to mention that the X-ray data provides information on the damaged DNA region, whereas PELDOR provides complementary information on changes in the global structure.

The bending of the DNA helix near the damaged nucleotide from PELDOR provides new information about the mechanism for recognition of damage by DNA repair enzymes. Detection of the bending allows one to understand why enzymes that slide along the DNA strand would stop at the damaged sites to repair them. The data is also important for understanding other enzymes that search for specific DNA sites.

In our opinion, the data obtained from PELDOR significantly contributes to the investigation of the structure and properties of DNA and RNA. This method opens new perspectives for studying complex nonlinear structures, interactions between polynucleotides and enzymes, proteins, and membranes. The potential of PELDOR as a method for structural studies will undoubtedly increase with the development of pulse EPR spectroscopy.

References

1. Schiemann O, Prisner TF (2007) Long-range distance determinations in biomacromolecules by EPR spectroscopy. Q Rev Biophys 40(1):1–53. https://doi.org/10.1017/S003358350700460x
2. Jeschke G, Polyhach Y (2007) Distance measurements on spin-labelled biomacromolecules by pulsed electron paramagnetic resonance. Phys Chem Chem Phys 9(16):1895–1910. https://doi.org/10.1039/B614920k
3. Tsvetkov YD, Milov AD, Maryasov AG (2008) Pulse electron-electron double resonance (PELDOR) as nanometre range EPR spectroscopy. Usp Khim 77(6):515–550
4. Sowa GZ, Qin PZ (2008) Site-directed spin labeling studies on nucleic acid structure and dynamics. Prog Nucleic Acid Re 82:147–197. https://doi.org/10.1016/S0079-6603(08)00005-6
5. Schiemann O (2009) Mapping global folds of oligonucleotides by pulsed electron-electron double resonance. Meth Enzymol 469: Biophysical, chemical, and functional probes of RNA structure, interactions and folding, Pt B 469:329–351. https://doi.org/10.1016/s0076-6879(09)69016-9

6. Reginsson GW, Schiemann O (2011) Studying bimolecular complexes with pulsed electron-electron double resonance spectroscopy. Biochem Soc T 39:128–139. https://doi.org/10.1042/Bst0390128
7. Reginsson GW, Schiemann O (2011) Pulsed electron-electron double resonance: beyond nanometre distance measurements on biomacromolecules. Biochem J 434:353–363. https://doi.org/10.1042/Bj20101871
8. Schiemann O, Weber A, Edwards TE, Prisner TF, Sigurdsson ST (2003) Nanometer distance measurements on RNA using PELDOR. J Am Chem Soc 125(12):3434–3435
9. Kolhe P, Amend E, Singh SK (2010) Impact of freezing on pH of buffered solutions and consequences for monoclonal antibody aggregation. Biotechnol Prog 26(3):727–733. https://doi.org/10.1002/btpr.377
10. Ward R, Keeble DJ, El-Mkami H, Norman DG (2007) Distance determination in heterogeneous DNA model systems by pulsed EPR. ChemBioChem 8(16):1957–1964. https://doi.org/10.1002/cbic.200700245
11. Schiemann O, Piton N, Mu YG, Stock G, Engels JW, Prisner TF (2004) A PELDOR-based nanometer distance ruler for oligonucleotides. J Am Chem Soc 126(18):5722–5729. https://doi.org/10.1021/ja0393877
12. Piton N, Mu YG, Stock G, Prisner TF, Schiemann O, Engels JW (2007) Base-specific spin-labeling of RNA for structure determination. Nucleic Acids Res 35(9):3128–3143
13. Cai Q, Kusnetzow AK, Hubbell WL, Haworth IS, Gacho GPC, Van Eps N, Hideg K, Chambers EJ, Qin PZ (2006) Site-directed spin labeling measurements of nanometer distances in nucleic acids using a sequence-independent nitroxide probe. Nucleic Acids Res 34 (17):4722–4730. https://doi.org/10.1093/nar/gkl546
14. Cai Q, Kusnetzow AK, Hideg K, Price EA, Haworth IS, Qin PZ (2007) Nanometer distance measurements in RNA using site-directed spin Labeling. Biophys J 93(6):2110–2117. https://doi.org/10.1529/biophysj.107.109439
15. Shevelev GY, Krumkacheva OA, Lomzov AA, Kuzhelev AA, Trukhin DV, Rogozhnikova OY, Tormyshev VM, Pyshnyi DV, Fedin MV, Bagryanskaya EG (2015) Triarylmethyl labels: toward improving the accuracy of EPR nanoscale distance measurements in DNAs. J Phys Chem B 119(43):13641–13648
16. Shevelev GY, Krumkacheva OA, Lomzov AA, Kuzhelev AA, Rogozhnikova OY, Trukhin DV, Troitskaya TI, Tormyshev VM, Fedin MV, Pyshnyi DV, Bagryanskaya EG (2014) Physiological-temperature distance measurement in nucleic acid using triarylmethyl-based spin labels and pulsed dipolar EPR spectroscopy. J Am Chem Soc 136 (28):9874–9877. https://doi.org/10.1021/ja505122n
17. Raitsimring AM, Gunanathan C, Potapov A, Efremenko I, Martin JML, Milstein D, Goldfarb D (2007) Gd^{3+} complexes as potential spin labels for high field pulsed EPR distance measurements. J Am Chem Soc 129(46):14138–14139. https://doi.org/10.1021/ja075544g
18. Potapov A, Song Y, Meade TJ, Goldfarb D, Astashkin AV, Raitsimring A (2010) Distance measurements in model bis-Gd(III) complexes with flexible "bridge". Emulation of biological molecules having flexible structure with Gd(III) labels attached. J Magn Reson 205 (1):38–49
19. Song Y, Meade TJ, Astashkin AV, Klein EL, Enemark JH, Raitsimring A (2011) Pulsed dipolar spectroscopy distance measurements in biomacromolecules labeled with Gd(III) markers. J Magn Reson 210(1):59–68. https://doi.org/10.1016/j.jmr.2011.02.010
20. Yang ZY, Kise D, Saxena S (2010) An approach towards the measurement of nanometer range distances based on Cu^{2+} Ions and ESR. J Phys Chem B 114(18):6165–6174. https://doi.org/10.1021/jp911637s
21. Sicoli G, Mathis G, Delalande O, Boulard Y, Gasparutto D, Gambarelli S (2008) Double electron-electron resonance (DEER): a convenient method to probe DNA conformational changes. Angew Chem Int Edit 47(4):735–737. https://doi.org/10.1002/anie.200704133
22. Yu H, Mu YG, Nordenskiold L, Stock G (2008) Influence of nitroxide spin labels on RNA structure: a molecular dynamics simulation study. J Chem Theory Comput 4(10):1781–1787. https://doi.org/10.1021/ct800266e

23. Romainczyk O, Endeward B, Prisner TF, Engels JW (2011) The RNA-DNA hybrid structure determined by EPR, CD and RNase H1. Mol BioSyst 7(4):1050–1052. https://doi.org/10.1039/c0mb00258e
24. Savitsky A, Dubinskii AA, Flores M, Lubitz W, Mobius K (2007) Orientation-resolving pulsed electron dipolar high-field EPR spectroscopy on disordered solids: I. Structure of spin-correlated radical pairs in bacterial photosynthetic reaction centers. J Phys Chem B 111 (22):6245–6262. https://doi.org/10.1021/jp070016c
25. Marko A, Margraf D, Yu H, Mu Y, Stock G, Prisner T (2009) Molecular orientation studies by pulsed electron-electron double resonance experiments. J Chem Phys 130(6):064102. https://doi.org/10.1063/1.3073040
26. Marko A, Margraf D, Cekan P, Sigurdsson ST, Schiemann O, Prisner TF (2010) Analytical method to determine the orientation of rigid spin labels in DNA. Phys Rev E 81(2):021911. https://doi.org/10.1103/PhysRevE.81.021911
27. Schiemann O, Cekan P, Margraf D, Prisner TF, Sigurdsson ST (2009) Relative orientation of rigid nitroxides by PELDOR: beyond distance measurements in nucleic acids. Angew Chem Int Edit 48(18):3292–3295
28. Barhate N, Cekan P, Massey AP, Sigurdsson ST (2007) A nucleoside that contains a rigid nitroxide spin label: a fluorophore in disguise. Angew Chem Int Edit 46(15):2655–2658
29. Gophane DB, Endeward B, Prisner TF, Sigurdsson ST (2014) Conformationally restricted isoindoline-derived spin labels in duplex DNA: distances and rotational flexibility by pulsed electron-electron double resonance spectroscopy. Chem-Eur J 20(48):15913–15919. https://doi.org/10.1002/chem.201403726
30. Tkach I, Halbmair K, Hobartner C, Bennati M (2014) High-frequency 263 GHz PELDOR. Appl Magn Reson 45(10):969–979. https://doi.org/10.1007/s00723-014-0581-z
31. Hagerman PJ (1988) Flexibility of DNA. Annu Rev Biophys Bio 17:265–286
32. Gore J, Bryant Z, Nollmann M, Le MU, Cozzarelli NR, Bustamante C (2006) DNA overwinds when stretched. Nature 442(7104):836–839. https://doi.org/10.1038/nature04974
33. Marko JF (1997) Stretching must twist DNA. Europhys Lett 38(3):183–188. https://doi.org/10.1209/epl/i1997-00223-5
34. Mathew-Fenn RS, Das R, Harbury PAB (2008) Remeasuring the double helix. Science 322 (5900):446–449. https://doi.org/10.1126/science.1158881
35. Marko A, Denysenkov V, Margraft D, Cekan P, Schiemann O, Sigurdsson ST, Prisner TF (2011) Conformational flexibility of DNA. J Am Chem Soc 133(34):13375–13379
36. Sicoli G, Wachowius F, Bennati M, Hobartner C (2010) Probing secondary structures of spin-labeled RNA by pulsed EPR spectroscopy. Angew Chem Int Edit 49(36):6443–6447. https://doi.org/10.1002/anie.201000713
37. Krstic I, Frolow O, Sezer D, Endeward B, Weigand JE, Suess B, Engels JW, Prisner TF (2010) PELDOR spectroscopy reveals preorganization of the neomycin-responsive riboswitch tertiary structure. J Am Chem Soc 132(5):1454–1455. https://doi.org/10.1021/ja9077914
38. Grytz CM, Marko A, Cekan P, Sigurdsson ST, Prisner TF (2016) Flexibility and conformation of the cocaine aptamer studied by PELDOR. Phys Chem Chem Phys 18 (4):2993–3002. https://doi.org/10.1039/c5cp06158j
39. Kim NK, Bowman MK, DeRose VJ (2010) Precise mapping of RNA tertiary structure via nanometer distance measurements with double electron-electron resonance spectroscopy. J Am Chem Soc 132(26):8882–8884. https://doi.org/10.1021/ja101317g
40. Zhang XJ, Tung CS, Sowa GZ, Hatmal MM, Haworth IS, Qin PZ (2012) Global structure of a three-way junction in a Phi29 packaging RNA dimer determined using site-directed spin labeling. J Am Chem Soc 134(5):2644–2652. https://doi.org/10.1021/ja2093647
41. Freeman ADJ, Ward R, El Mkami H, Lilley DMJ, Norman DG (2011) Analysis of conformational changes in the DNA junction-resolving enzyme T7 endonuclease I on binding a four-way junction using EPR. Biochemistry-Us 50(46):9963–9972. https://doi.org/10.1021/bi2011898

42. Danielsson J, Inomata K, Murayama S, Tochio H, Lang LS, Shirakawa M, Oliveberg M (2013) Pruning the ALS-associated protein SOD1 for in-cell NMR. J Am Chem Soc 135 (28):10266–10269. https://doi.org/10.1021/ja404425r
43. Banci L, Barbieri L, Bertini I, Luchinat E, Secci E, Zhao YG, Aricescu AR (2013) Atomic-resolution monitoring of protein maturation in live human cells by NMR. Nat Chem Biol 9(5):297–300. https://doi.org/10.1038/Nchembio.1202
44. Sakai T, Tochio H, Tenno T, Ito Y, Kokubo T, Hiroaki H, Shirakawa M (2006) In-cell NMR spectroscopy of proteins inside *Xenopus laevis* oocytes. J Biomol NMR 36(3):179–188. https://doi.org/10.1007/s10858-006-9079-9
45. Ogino S, Kubo S, Umemoto R, Huang SX, Nishida N, Shimada I (2009) Observation of NMR signals from proteins introduced into living mammalian cells by reversible membrane permeabilization using a pore-forming toxin, streptolysin O. J Am Chem Soc 131(31):10834–10835. https://doi.org/10.1021/ja904407w
46. Azarkh M, Singh V, Okle O, Seemann IT, Dietrich DR, Hartig JS, Drescher M (2013) Site-directed spin-labeling of nucleotides and the use of in-cell EPR to determine long-range distances in a biologically relevant environment. Nat Protoc 8(1):131–147. https://doi.org/10.1038/nprot.2012.136
47. Azarkh M, Okle O, Eyring P, Dietrich DR, Drescher M (2011) Evaluation of spin labels for in-cell EPR by analysis of nitroxide reduction in cell extract of *Xenopus laevis* oocytes. J Magn Reson 212(2):450–454. https://doi.org/10.1016/j.jmr.2011.07.014
48. Krstic I, Hansel R, Romainczyk O, Engels JW, Dotsch V, Prisner TF (2011) Long-range distance measurements on nucleic acids in cells by pulsed EPR spectroscopy. Angew Chem Int Edit 50(22):5070–5074. https://doi.org/10.1002/anie.201100886
49. Igarashi R, Sakai T, Hara H, Tenno T, Tanaka T, Tochio H, Shirakawa M (2010) Distance determination in proteins inside *Xenopus laevis* oocytes by double electron-electron resonance experiments. J Am Chem Soc 132(24):8228–8229. https://doi.org/10.1021/ja906104e
50. Bowman MK, Maryasov AG, Kim N, DeRose VJ (2004) Visualization of distance distribution from pulsed double electron-electron resonance data. Appl Magn Reson 26(1–2):23–39. https://doi.org/10.1007/Bf03166560
51. Nakata M, Zanchetta G, Chapman BD, Jones CD, Cross JO, Pindak R, Bellini T, Clark NA (2007) End-to-end stacking and liquid crystal condensation of 6-to 20-base pair DNA duplexes. Science 318(5854):1276–1279. https://doi.org/10.1126/science.1143826
52. Duchardt-Ferner E, Weigand JE, Ohlenschlager O, Schtnidtke SR, Suess B, Wohnert J (2010) Highly modular structure and ligand binding by conformational capture in a minimalistic riboswitch. Angew Chem Int Edit 49(35):6216–6219. https://doi.org/10.1002/anie.201001339
53. Nozinovic S, Furtig B, Jonker HRA, Richter C, Schwalbe H (2010) High-resolution NMR structure of an RNA model system: the 14-mer cUUCGg tetraloop hairpin RNA. Nucleic Acids Res 38(2):683–694. https://doi.org/10.1093/nar/gkp956
54. Bochman ML, Paeschke K, Zakian VA (2012) DNA secondary structures: stability and function of G-quadruplex structures. Nat Rev Genet 13(11):770–780. https://doi.org/10.1038/nrg3296
55. Singh V, Azarkh M, Exner TE, Hartig JS, Drescher M (2009) Human telomeric quadruplex conformations studied by pulsed EPR. Angew Chem Int Edit 48(51):9728–9730. https://doi.org/10.1002/anie.200902146
56. Wang Y, Patel DJ (1993) Solution structure of the human telomeric repeat d[$AG_3(T_2AG_3)_3$] G-tetraplex. Structure 1(4):263–282. https://doi.org/10.1016/0969-2126(93)90015-9
57. Phan AT, Kuryavyi V, Luu KN, Patel DJ (2007) Structure of two intramolecular G-quadruplexes formed by natural human telomere sequences in K^+ solution. Nucleic Acids Res 35(19):6517–6525. https://doi.org/10.1093/nar/gkm706
58. Parkinson GN, Lee MPH, Neidle S (2002) Crystal structure of parallel quadruplexes from human telomeric DNA. Nature 417(6891):876–880. https://doi.org/10.1038/nature755

59. Azarkh M, Singh V, Okle O, Dietrich DR, Hartig JS, Drescher M (2012) Intracellular conformations of human telomeric quadruplexes studied by electron paramagnetic resonance spectroscopy. ChemPhysChem 13(6):1444–1447. https://doi.org/10.1002/cphc.201100980
60. Shiokawa K, Tashiro K, Yamana K, Sameshima M (1987) Electron-microscopic studies of giant nucleus-like structure formed by lambda DNA introduced into the cytoplasm of *Xenopus laevis* fertilized-eggs and embryos. Cell Differ Dev 20(4):253–261. https://doi.org/10.1016/0045-6039(87)90470-2
61. Singh V, Azarkh M, Drescher M, Hartig JS (2012) Conformations of individual quadruplex units studied in the context of extended human telomeric DNA. Chem Commun 48(66):8258–8260. https://doi.org/10.1039/c2cc32012f
62. Martorana A, Bellapadrona G, Feintuch A, Di Gregorio E, Aime S, Goldfarb D (2014) Probing protein conformation in cells by EPR distance measurements using Gd^{3+} spin labeling. J Am Chem Soc 136(38):13458–13465. https://doi.org/10.1021/ja5079392
63. Goldfarb D (2014) Gd^{3+} spin labeling for distance measurements by pulse EPR spectroscopy. Phys Chem Chem Phys 16(21):9685–9699. https://doi.org/10.1039/c3cp53822b
64. Thonon D, Jacques V, Desreux JF (2007) A gadolinium triacetic monoamide DOTA derivative with a methanethiosulfonate anchor group. Relaxivity properties and conjug. Contrast Media Mol I 2(1):24–34. https://doi.org/10.1002/cmmi.121
65. Rossi L, Serafini S, Pierige F, Antonelli A, Cerasi A, Fraternale A, Chiarantini L, Magnani M (2005) Erythrocyte-based drug delivery. Expert Opin Drug Deliv 2(2):311–322. https://doi.org/10.1517/17425247.2.2.311
66. Markov DE, Boeve H, Gleich B, Borgert J, Antonelli A, Sfara C, Magnani M (2010) Human erythrocytes as nanoparticle carriers for magnetic particle imaging. Phys Med Biol 55 (21):6461–6473. https://doi.org/10.1088/0031-9155/55/21/008
67. Hara H, Tenno T, Shirakawa M (2007) Distance determination in human ubiquitin by pulsed double electron-electron resonance and double quantum coherence ESR methods. J Magn Reson 184(1):78–84. https://doi.org/10.1016/j.jmr.2006.09.017
68. Sicoli G, Mathis G, Aci-Seche S, Saint-Pierre C, Boulard Y, Gasparutto D, Gambarelli S (2009) Lesion-induced DNA weak structural changes detected by pulsed EPR spectroscopy combined with site-directed spin labelling. Nucleic Acids Res 37(10):3165–3176. https://doi.org/10.1093/nar/gkp165
69. Kuznetsov NA, Milov AD, Koval VV, Samoilova RI, Grishin YA, Knorre DG, Tsvetkov YD, Fedorova OS, Dzuba SA (2009) PELDOR study of conformations of double-spin-labeled single- and double-stranded DNA with non-nucleotide inserts. Phys Chem Chem Phys 11 (31):6826–6832. https://doi.org/10.1039/b904873a
70. Kuznetsov NA, Milov AD, Isaev NP, Vorobjev YN, Koval VV, Dzuba SA, Fedorova OS, Tsvetkov YD (2011) PELDOR analysis of enzyme-induced structural changes in damaged DNA duplexes. Mol BioSyst 7(9):2670–2680. https://doi.org/10.1039/c1mb05189j
71. Wunnicke D, Ding P, Seela F, Steinhoff HJ (2012) Site-directed spin labeling of DNA reveals mismatch-induced nanometer distance changes between flanking nucleotides. J Phys Chem B 116(14):4118–4123. https://doi.org/10.1021/jp212421c
72. Flaender M, Sicoli G, Aci-Seche S, Reignier T, Maurel V, Saint-Pierre C, Boulard Y, Gambarelli S, Gasparutto D (2011) A triple spin-labeling strategy coupled with DEER analysis to detect DNA modifications and enzymatic repair. ChemBioChem 12(17):2560–2563. https://doi.org/10.1002/cbic.201100550
73. Jeschke G, Chechik V, Ionita P, Godt A, Zimmermann H, Banham J, Timmel CR, Hilger D, Jung H (2006) DeerAnalysis2006—a comprehensive software package for analyzing pulsed ELDOR data. Appl Magn Reson 30(3–4):473–498. https://doi.org/10.1007/Bf03166213
74. Jeschke G, Sajid M, Schulte M, Godt A (2009) Three-spin correlations in double electron-electron resonance. Phys Chem Chem Phys 11(31):6580–6591
75. Takeuchi M, Lillis R, Demple B, Takeshita M (1994) Interactions of *Escherichia coli* endonuclease-IV and exonuclease-III with abasic sites in DNA. J Biol Chem 269(34):21907–21914

76. Banerjee A, Santos WL, Verdine GL (2006) Structure of a DNA glycosylase searching for lesions. Science 311(5764):1153–1157. https://doi.org/10.1126/science.1120288
77. Qi Y, Spong MC, Nam K, Karplus M, Verdine GL (2010) Entrapment and structure of an extrahelical guanine attempting to enter the active site of a bacterial DNA glycosylase, MutM. J Biol Chem 285(2):1468–1478. https://doi.org/10.1074/jbc.M109.069799
78. Gilboa R, Zharkov DO, Golan G, Fernandes AS, Gerchman SE, Matz E, Kycia JH, Grollman AP, Shoham G (2002) Structure of formamidopyrimidine-DNA glycosylase covalently complexed to DNA. J Biol Chem 277(22):19811–19816. https://doi.org/10.1074/jbc.M202058200

Chapter 8
PELDOR in Photo- and Radiation Chemistry

Abstract Early applications of PELDOR or DEER spectroscopy involved photochemistry and radiation chemistry. The atoms and molecules that absorbed the initial radiation; the radical pairs and/or excited states produced in that primary interaction of radiation with matter; and the subsequent chemical reactions are all long-standing issues that could be addressed by accurate measurements of pair distribution functions between the free radical products. The same issues arise in biological photosynthesis. The movement of electrons and holes during the entire sequence of reactions provided a clear picture of the sequence of reactions. Some of the initial applications of pulsed EPR methods, in particular ESE, to molecular systems focused on radiation chemistry and photochemistry (Salikhov et al in Electron spin echo and its applications. Nauka, Novosibirsk, 1976 [1]; Salikhov and Tsvetkov et al in Time domain electron spin resonance. John Wiley, New York, 1979 [2]; Bowman et al in Applications of EPR in radiation research. Springer, Heidelberg, pp 581–627, 2014 [3]). PELDOR emerged as a method focusing on the spatial distributions of products.

8.1 Radical Pairs Formed by Photolysis

8.1.1 Hydroquinone in Sulfuric Acid

Many important physicochemical processes form paramagnetic centers in pairs. For instance, the ultraviolet, or UV, photolysis of solids often produces an electron which becomes trapped by acceptors or the medium after travelling some distance. The holes resulting from loss of these electrons undergo analogous motion and trapping. Reactions of products with each other also occur. The distance between the pairs of paramagnetic products is described by a distance distribution function $F(r)$ and can be studied by PELDOR. If $F(r)$ is rather broad, the PELDOR signal has no visible oscillations. Nevertheless, information on the distribution of radical pairs can be obtained from analysis of the PELDOR time trace.

© Springer Nature Switzerland AG 2019

Y. D. Tsvetkov et al., *Pulsed Electron–Electron Double Resonance*,
https://doi.org/10.1007/978-3-030-05372-7_8

The pioneering PELDOR study [4] investigated radical pairs, composed of a hydrogen atom and a hydroquinone radical, trapped in frozen 8 M aqueous H_2SO_4 or D_2SO_4 glasses containing dissolved hydroquinone. The paramagnetic centers are formed by photoionization of the hydroquinone $Ar(OH)_2$ by UV light at 77 K:

$$\begin{array}{c} Ar(OH)_2 \xrightarrow{UV} Ar(OH)_2^{\cdot+} + e^{\cdot-} \\ e^{\cdot-} + H_3O^+ \rightarrow H^{\cdot} + H_2O \\ \hline Ar(OH)_2 + H_3O^+ \xrightarrow{UV} Ar(OH)_2^{\cdot+} + H^{\cdot} + H_2O \end{array}$$

The observed distance distribution is governed by two factors: (1) the free path length of an electron prior to trapping by H_3O^+; and (2) the diffusion path length of hydrogen atoms before trapping in the matrix. In the PELDOR experiments, the hydrogen atoms were the A spins and produced the spin echo signals. The pump pulse was applied at the EPR frequency of the hydroquinone radical ions $Ar(OH)_2^{\cdot+}$, which were the B spins, Fig. 8.1a. The PELDOR time trace $V(T)$ from the hydrogen atoms depends on the concentration of hydroquinone radical ions. This means that each hydrogen atom has dipolar interactions not just with its partner, or geminate, hydroquinone radical, but also with other hydroquinone radicals.

The PELDOR time trace can be separated into $V(T) = V_{INTER}V_{INTRA}$ because the hydroquinone radicals are distributed uniformly throughout the matrix, Eq. 1.13. The V_{INTRA} was determined from the dependence of $V(T)$ on the concentration of hydroquinone radicals, and was then used to determine $F(r)$, approximating $\Phi(T)$, Eq. 1.11, in those early days of PELDOR, as a step function. The PELDOR data gave a distribution function with a maximum at 4.5 nm, Fig. 8.1. Integration of $F(r)$ reveals that 63% of the radicals are within 8.5 nm of their geminate radical partner and nearly all are within 10 nm. The actual shape of $F(r)$ is difficult to determine at the longer distances because the distant radical pairs have little impact on $V(T)$.

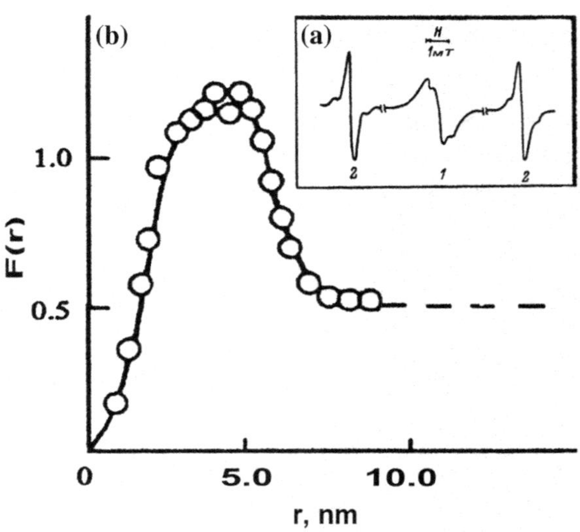

Fig. 8.1 UV-irradiated, frozen hydroquinone solutions in 8 M H_2SO_4, **a** EPR spectrum of paramagnetic centers formed in this system: *1* hydroquinone radicals, B spins and *2* hydrogen atoms, A spins; **b** the $F(r)$ function calculated from the PELDOR data. Reprinted from Milov et al. [4]

8.1.2 Aromatic Molecules in Hydrocarbons

The UV photolysis of aromatic compounds in hydrocarbon solutions at 77 K produces trapped solvent radicals. The mechanism for their formation involves absorption of two photons by an aromatic molecule with subsequent transfer of energy or charge to a nearby hydrocarbon solvent molecule followed by radical decomposition [5].

UV photolysis of naphthalene in glassy decalin at 77 K by light with $\lambda >$ 290 nm produces radicals whose EPR spectrum consists of six non-equidistant lines [6]. This spectrum belongs to radicals formed by abstraction of a hydrogen atom from a decalin molecule. The PELDOR time trace $V(T)$, at low concentrations of decalin radicals, reaches a limiting value V_p rather quickly, by $T \sim 100$ ns. Such behavior is characteristic of radicals distributed in clusters. The $V(T)$ of nitroxyl biradicals **3-9** and **3-10** have the same shape.

The PELDOR measurements used a bimodal resonator Sect. 2.5 whose modes were independently tuned to different lines of the decalin radical for ω_A and ω_B. This simplified the study of the V_p dependence on p_B. Changing the B_0 value moved the EPR spectrum relative to the pump and observe frequencies while the pulse amplitudes and durations remained constant. The V_p is a linear function of p_B with a slope that is independent of radical concentration or the extent of photolysis, Fig. 8.2. The relation $V_p \approx 1 - (N - 1)p_B$, Eq. 1.26, was first derived for analysis

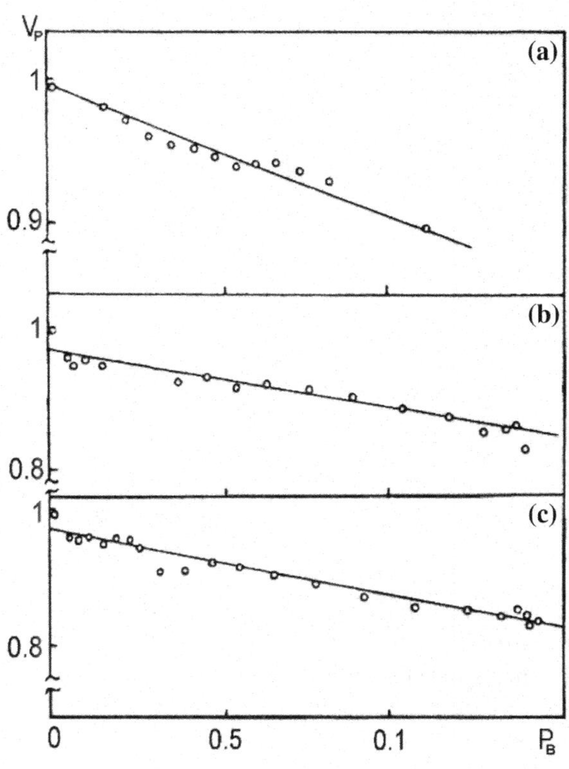

Fig. 8.2 Plots of the limiting V_p values versus p_B: **a** biradical **3.9**, **b** decalin radicals at 5.6×10^{17} cm^{-3} and **c** decalin radicals at 3×10^{17} cm^{-3}. Reprinted from Milov et al. [6] with permission of Elsevier B.V., copyright 1984

of this data [6]. The slope of the experimental straight line gave the number of radicals per cluster as $N = 1.85 \pm 0.2$. Thus, sensitized low-temperature photolysis of decalin produces stabilized, decalin radicals pairs. A limiting distance of 2.5 nm between radicals in the pair was estimated.

8.2 Radical Pairs in Photosynthetic Systems

Photosynthesis is one of the most important physicochemical processes. It converts solar light energy into electrochemical energy and is the basis for life on earth. The photosynthetic reactions involve energy, electron and proton transfers, with radical ions and free radicals as direct participants or by-products. Information on distances between these paramagnetic centers and their mutual orientation is of considerable importance in photosynthesis research since it provides a critical connection between structure and function. The vast majority of photosynthetic studies by PELDOR were carried out on Photosystem II (PSII) of green plants. The structural context for the photosynthetic reactions was known from X-ray diffraction studies [7] and a general outline of the reactions was known from extensive biophysical and spectroscopic studies. But the choreography of the movement of electrons, protons, molecules and proteins, as energy was captured and stored, was sometimes the focus of intense speculation. The overall scheme in PSII is roughly sketched below:

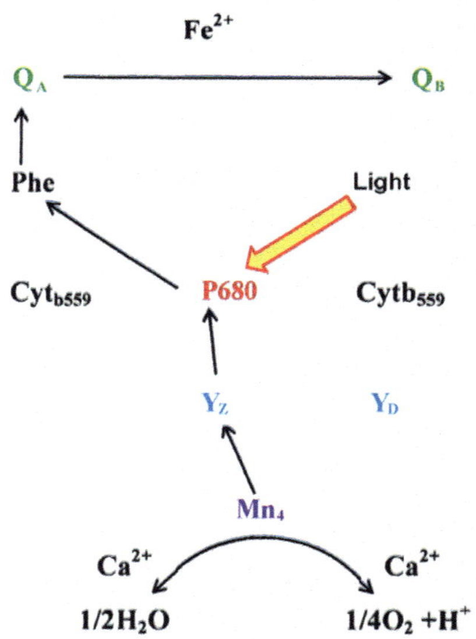

At the center of PSII is the primary electron donor P680, a chlorophyll *a* dimer with a distinctive optical absorption maximum at 680 nm. Light is absorbed by antenna pigments that funnel the excitation energy to P680, initiating a series of electron transfers in PSII. An electron from the excited P680 is relayed, *via* an intermediate acceptor pheophytin Phe, to the primary electron acceptor, quinone Q_A, and later on to the secondary acceptor, quinone Q_B. The electron vacancy or hole on P680 is rapidly filled by a sequence of redox reactions involving tyrosine radicals Y_Z and Y_D, and a cluster of manganese ions Mn_4. The Mn_4 carries out the crucial steps of water splitting and formation of oxygen. Each electron can be followed through the series of transient paramagnetic centers detected by EPR spectroscopy. These centers, along with systems that model various properties of the reaction center of PSII, has been studied extensively by magnetic resonance [8, 9].

8.2.1 Stable Radical Pairs

PELDOR and other pulsed dipolar methods were applied to several relatively-stable radical pairs in PSII. At room temperature, the oxidized $Y_Z^{\cdot +}$ has a short lifetime and can barely be detected by EPR spectroscopy. However, oxidized $Y_D^{\cdot +}$ is stable in the dark and is readily detected. The EPR spectrum of $Y_D^{\cdot +}$ is a single, narrow line with $g \approx 2$. The Mn_4 cluster has five states (S_0–S_4); each state has a different EPR spectrum. In addition, removal of a calcium ion (Ca) from the Mn_4 cluster leaves it redox active but unable to release oxygen. A review of stable paramagnetic centers in PSII studied by PELDOR and other EPR methods can be found in [10].

The oxygen-evolving and the Ca-depleted PSII systems were studied [11]. The EPR spectra of the S_2 and S_3 states of Mn_4 are quite different from those of $Y_D^{\cdot +}$ which made it possible to measure the dipolar coupling between these species by PELDOR. The pump pulse was applied to $Y_D^{\cdot +}$ and the spin echo signal was produced from Mn_4 and shows clear modulation, Fig. 8.3. The distance between the unpaired electron spin in the S_2 state and $Y_D^{\cdot +}$ is $r = 2.7\,\text{nm}$ in oxygen-evolving and in Ca-depleted PSII, Table 8.1. The distance between the spins of the S_3 state of Ca-depleted PSII and $Y_D^{\cdot +}$ is slightly larger at 3.0 nm, and even longer for the S_o state of the Mn_4 cluster [11]. The distances between the $Y_D^{\cdot +}$ cation and the spin of Mn_4 are different because the spin of Mn_4 is localized on different ions of the cluster in its different S states.

PELDOR experiments with other stable radical pairs in PSII are difficult at X-band because their EPR spectra have strong overlap. It is often impossible to select ω_A and ω_B for selective mw excitation of individual centers. But X-band EPR experiments were possible between a number of centers, Table 8.1 [11–16].

The reaction centers of PSII are embedded in specific structures of the chloroplast membrane called thylakoids. The donor and acceptor regions of PSII are localized on opposite sides of the thylakoid membrane. Their orientation relative to the membrane surface was studied by the first PELDOR orientation measurements [17]. A Mylar sheet was coated with a layer of alligned PSII and the sheet was oriented

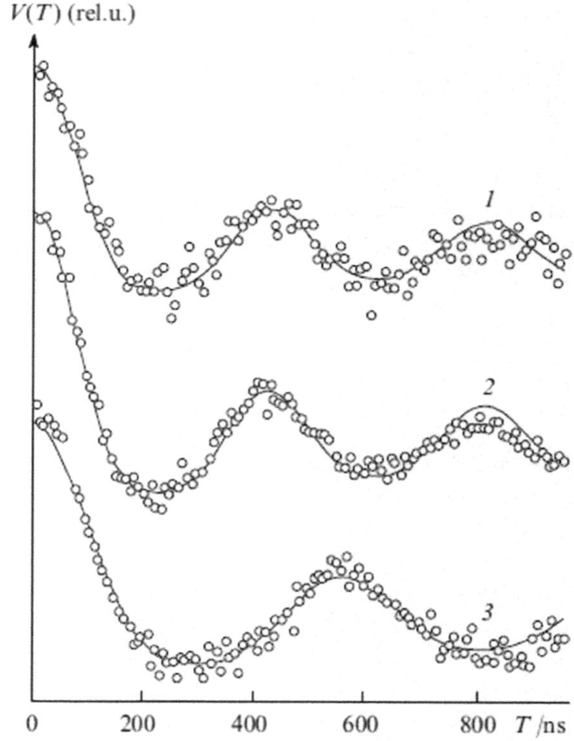

Fig. 8.3 The $V(T)$ of pairs of paramagnetic centers in PSII: *1* oxygen-producing PSII with active Mn_4 clusters in S_2; *2* Ca-depleted PSII with Mn_4 in S_2; and *3* Ca-depleted PSII with Mn_4 in S_3; *points* denote experimental data and *solid lines* denote calculations. Reproduced from Tsvetkov et al. [39]

Table 8.1 Distances between paramagnetic centers in PSII determined by PELDOR

Pair of centers	Distance, nm	Reference
Y_D—$Mn_4(S_2)$	2.71 ± 0.02	[11]
Y_D—$Mn_4(S_3)$	2.97 ± 0.02	[11]
Y_D—$Mn_4(S_0)$	3.40 ± 0.05	[12]
Q_A—Chl_Z	3.4 ± 0.1	[13]
Q_A—Y_Z	3.4 ± 0.05	[14]
Q_A—Y_Z	3.45 ± 0.1	[15]
Q_A—Cyt_{b559}	4.0 ± 0.1	[16]

relative to the external magnetic field B_0. PELDOR time traces of the radical pair consisting of $Y_D^{\cdot +}$ and the S_2 state of Mn_4 were measured. The vector connecting the two centers makes a 20 ±2° angle with the normal to the thylakoid membrane surface.

Radical pairs have also been produced by spin labeling proteins of the photosynthetic apparatus. This approach was used to study the protein component of light-harvesting complex LHCIIb, which contains chlorophylls *a* and *b* and is present in both PSI and PSII. Spin labels were introduced pairwise into the protein and the distance distribution $F(r)$ between pairs were determined [18]. The $F(r)$ for all labels near the N-terminus of the protein appear to be bimodal, corresponding to two different conformations. The two conformers were assigned to LHCIIb in the two different photosystems, namely, PSI and PSII [18]. The N-terminal domain is involved in light flux regulation, suggesting a role for conformational changes.

8.2.2 Transient Radical Pairs

PELDOR studies on stable radical pairs in 8.2.1 were made at X-band. Higher EPR frequencies, where g-values can be resolved, provide additional opportunities to obtain the orientation of the donors, acceptors and cofactors in addition to distances. The most extensive high-frequency studies [19–21] focused on transient radical pairs of the reaction center (RC) of photosynthetic bacteria, Fig. 8.4.

A short-lived radical pair, $P_{865}{}^{\cdot +}Q_A{}^{\cdot -}$, Fig. 8.4, was produced by a laser pulse in frozen solutions of reaction centers from the photosynthetic bacteria *Rb. sphaeroides* and studied by W-band EPR spectroscopy at 95 GHz [19, 20]. Some of the reactions actually become faster at low temperature, so measurements were made near 150 K to reduce the radical pair lifetime. At W-band, the EPR spectra of these radical ions only partially overlap, which enables measurements at the g_{xx}, g_{yy}, g_{zz} of $Q_A{}^{\cdot -}$ and $P_{865}{}^{\cdot +}$ [22], so that their orientation within the pair as well as their

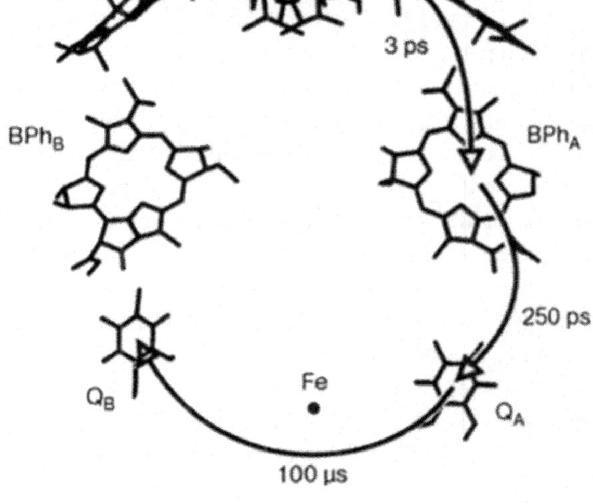

Fig. 8.4 X-ray structure of the cofactors in the reaction center of *Rhodobacter sphaeroides*; the cofactors P_{865} (bacteriochlorophyll *a* dimer), $BChl_A$ and $BChl_B$ (bacteriochlorophyll *a*), BPh_A and BPh_B (bacteriopheophytin *a*), Q_A and Q_B (ubiquinone-10), Fe (Fe^{2+} ion) are related by approximate C_2 symmetry of the RC proteins; yet, the light-induced electron transfer (ET) proceeds almost exclusively along the A branch; *Arrows* indicate the ET steps with their time constants. Reproduced from Schnegg et al. [20], with permission Springer-Verlag, copyright 2007

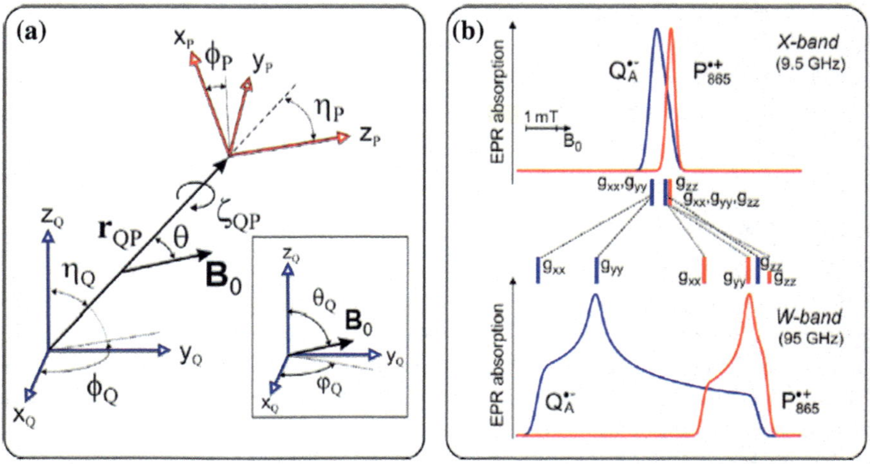

Fig. 8.5 **a** Geometry of the g-tensor frames $R_Q(x_Q, y_Q, z_Q)$ and $R_P(x_P, y_P, z_P)$ in terms of polar angles η_Q, ϕ_Q and η_P, ϕ_P and the dipolar vector r_{QP}; the angle ζ_{QP} is the relative tilt of the Q and P frames around the dipolar axis; **b** EPR absorption spectra of $P_{865}{}^{\cdot+}$ and $Q_A{}^{\cdot-}$ radicals at X-band (9.5 GHz) and W-band (95 GHz); the principal g-values of both radicals are indicated, with the maximum amplitude of each spectrum normalized. Reproduced from Schnegg et al. [20] with permission Springer Nature, copyright 2007

distance were determined. The orientation of g-axes within the radical pair is shown in Fig. 8.5.

The RC radical pair produced by light is born in the singlet state. Therefore, the EPR spectrum is spin-polarized and has adsorptive and emissive components from the spin-correlated radical pair (SCRP) mechanism [23–26] and somewhat different selection rules [27, 28].

Figure 8.6a shows the mw pulse scheme used for 4pPELDOR on pulsed laser-generated SCRPs. The frequency ω_A of the three observe π/2 pulses was adjusted to an EPR transition of the observed spins, either $P_{865}{}^{\cdot+}$ or $Q_A{}^{\cdot-}$. During the time period T, the mw π pump pulse at ω_B was applied to flip the partner spins. The ω_A and ω_B, Fig. 8.6a, correspond to the resonance conditions at constant field for the principal g-values in the normal field-swept EPR spectrum.

The observed spins dephase during the time τ between the first and second π/2 observe pulses, in part from the dipolar field of the partner radical. But here, the dipolar field is reversed during τ following the third observe pulse when the A spins are refocused into a spin echo. Reversing the dipolar field modulates the spin echo with the frequency of the dipolar interaction.

The SCRP state produces two different echo signals, an in-phase signal S_{-y} and an out-of-phase signal S_x [29]. Figure 8.6 shows a typical $V(T)$ slice for S_x. The complete set of slices, measured by stepping the mw frequency ω_B through the spectrum of the partner radical, contains modulation for the pump pulse at each position within the EPR spectrum of the partner radical. The PELDOR spectrum is

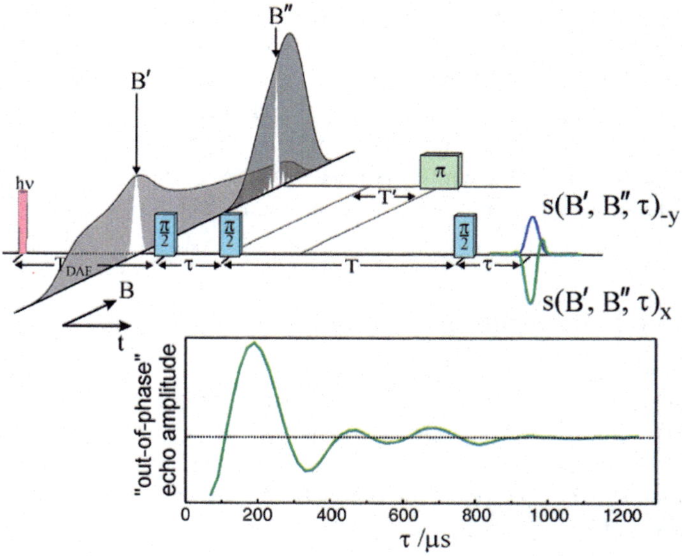

Fig. 8.6 *Upper:* Pulse sequence for the W-band PELDOR experiment on spin-correlated radical pairs generated by a laser flash in photosynthetic reaction centers: the thermally-equilibrated EPR spectra of both radicals, $P_{865}^{\cdot +}$ and $Q_A^{\cdot -}$, as well as the mw excitation bandwidths for typical mw pulse length settings are shown as a visual aid; *lower:* a representative example of a dipolar modulation echo-decay trace from the out-of-phase-detected echo versus τ is shown. Reproduced from Savitsky et al. [19] with permission American Chemical Society, copyright 2007

detected only for those $P_{865}^{\cdot +}$ or $Q_A^{\cdot -}$ radicals in a SCRP which are coupled by a dipolar interaction.

Information about the full 3D geometry of the radical pair, Fig. 8.5, is encoded in the modulation of these PELDOR spectra. Geometrical parameters for the $P_{865}^{\cdot +}Q_A^{\cdot -}$ radical pair were extracted by newly-developed methods [19, 20] and are compared to the ground-state $P_{865}^{\cdot +}Q_A^{\cdot -}$ radical pair, Table 8.2. No significant differences were seen, meaning that no substantial reorientation of P_{865} and Q_A occurs during the light-driven charge separation that produces the $P_{865}^{\cdot +}$ and $Q_A^{\cdot -}$ radical ion pair [7, 30].

Table 8.2 Angles and distances describing the geometry of the $P_{865}^{\cdot +}Q_A^{\cdot -}$ pair

P_{865} Q_A	η_Q^0	ϕ_Q^0	η_P^0	ϕ_P^0	ζ_{QP}^0	r, nm
Ground state pair[a]	97.4	246.2	66.2	79.5	49.5	2.84
PELDOR radical pair[b]	109.5	242	59	82	50	2.94

The angles are for the transition from the Q frame to the P frame, i.e., the dipolar vector direction is Q-P, Fig. 8.5a

[a]Reference pair: obtained from X-ray crystallographic data [19] of the dark state RC complemented by the W-band EPR single crystal investigation of $P_{865}^{\cdot +}$ [7]

[b]PELDOR radical pair: obtained from ESE-detected W-band EPR combined with PELDOR

The cofactor geometry in PSI was also studied by high-field PELDOR [31]. The short-lived ion-radical pair $P_{700}^{\cdot+}A_1^{\cdot-}$, where A_1 is phylloquinone, is analogous to $P_{865}^{\cdot+}Q_A^{\cdot-}$. The position and orientation of the reduced $A_1^{\cdot-}$ coincide with those of its parent A_1; again, no substantial orientation changes occur upon charge separation. On the other hand, several distinct orientations of the $P_{700}^{\cdot+}$ g-tensor axes were found and attributed to conformational substates of the $P_{700}^{\cdot+}$ radical ion with slightly different electron spin density distributions [31].

8.3 Spatial Distribution of Radicals from Radiolysis

The energy of hard radiation, e.g., α-, β- and γ-rays, and nuclear fission products, is deposited non-uniformly in liquids and solids, producing intermediates and reaction products, e.g., ions, electrons, atoms, and radicals, that are also spatially non-uniform. The regions containing products are known as tracks and spurs. These radical tracks and spurs were an early focus of pulse EPR and ESE in chemistry, e.g., see reviews [2, 3, 32].

Analysis of the PELDOR time trace $V(T)$ characterizes the number of spins in a group much better than classic spin echo methods. Even if there is a distribution in the number of spins in a group, PELDOR can give a weighted average or effective number of spins, Sects. 1.2.6 and 1.2.7. Consequently, one early application of PELDOR was to characterize the radical clusters in γ-irradiated cyclohexane, polyethylene and a number of organic acids [33–35].

Cyclohexyl radicals in crystalline cyclohexane, γ-irradiated at 77 K, produce a $V(T)$ similar to that of biradicals with a broad distance distribution [34]. There is an initial decay, in about 0.5 μs, to a limiting value V_p. The dependence of V_p on p_A and p_B was different from that of biradicals or radical pairs produced by sensitized photolysis, Sect. 8.1. The parameter N_{eff} depends on p_A, the probability or extent of excitation by observe pulses at ω_A for the cyclohexyl radicals while a biradical control **8-1** gave $N_{eff} = \sim 2.2$ independent of p_A, Fig. 8.7.

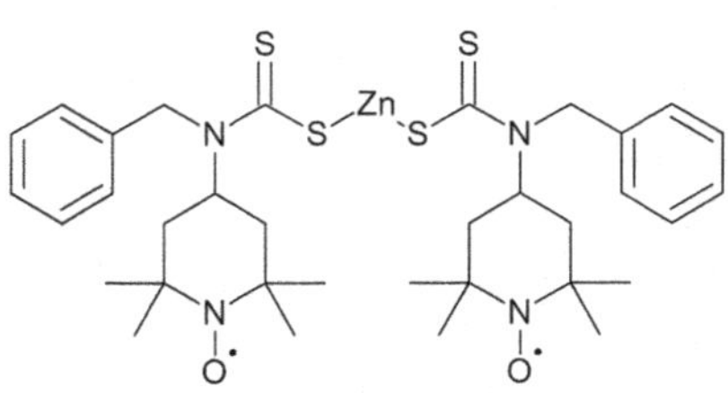

The dependence of N_{eff} on p_A in γ-irradiated cyclohexane is due to groups containing different numbers of radicals. Analysis of the change of V_p with p_A, Sects. 1.2.6, 1.2.7, and 1.5, gave some information about the number of radicals in

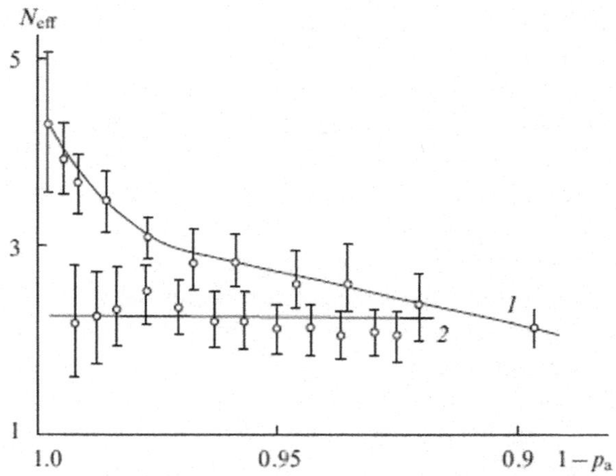

Fig. 8.7 Effective number of radicals, N_{eff}, versus excitation probability, p_A, by the observe pulses at ω_A: *1* for cyclohexyl radicals and *2* for nitroxyl biradicals **8-1**. Reproduced from Tsvetkov et al. [39]

a group. Most radicals, 75–90%, are distributed uniformly or in pairs, while the remaining 10–25% are in clusters containing three to several tens of radicals. The characteristic group size determined from the initial $V(T)$ is $r = \sim 2.5$ nm, which corresponds to a local radical density of $\sim 10^{20}$ cm^{-3}. High local concentrations like this are missed by two-pulse ESE because the dead time for ESE is greater than the short relaxation time at such radical densities. Thus, PELDOR gives much more detailed information about the spatial distribution of radicals in γ-irradiated systems than do ESE methods.

Alkyl radicals produced by γ-radiolysis of polyethylene and some organic acids gave similar results [35]. Irradiation and measurements were made at 77 K to minimize further radical reactions. Irradiated polyethylene, like irradiated cyclohexane, gave $V(T)$ that decay rapidly at short times, reaching a limiting value V_p, which depends on p_A and p_B. Careful analysis, Sect. 1.2.7, determined N_{eff} and the first and second moments of the distribution of number of radicals in groups [35]. Low-temperature radiolysis of polyethylene produces alkyl radicals in groups with an average of 2.4 radicals, second moment ≤ 6, and a radius of ≤ 2.5 nm.

PELDOR was used to probe the spatial distribution of radicals in DNA irradiated at 77 K to doses of 1.7–50 kGy by heavy-ion beams of 100 meV per nucleon ^{40}Ar ions having an LET of 300–400 keV/μm [36]. Such ions produce dense tracks of damage in their wake with extensive recombination of paramagnetic centers. At the doses used, the individual tracks are well separated and the samples had superimposable PELDOR spectra. The PELDOR time traces decayed smoothly over a range of p_B that varied by a factor of 12. At the smaller values of p_B, the signal intensity reached a limiting value that allowed determination of the number N of radicals contributing to the PELDOR time trace. The rate of decay to that limiting value gave the local concentration of those paramagnetic centers. The p_B was underestimated, slightly overestimating N and local concentration. This error is partly offset when the track radius is calculated. A global fit of all samples and data gave $C_{loc} \sim 13.5 \times 10^{19}$ radicals cm^{-3} and a track radius of 6.8 nm, consistent with the LET and radiation chemical yield.

Polycrystalline ammonium tartrate was studied by PELDOR after γ-, neutron- and 19.3 meV proton-irradiation [37] following initial spin echo measurements of instantaneous diffusion. This is the first reported case of PELDOR $V(T)$ in irradiated materials with pronounced modulation not caused by nuclear modulation (ESEEM) artifacts [38]. The resolved modulation was attributed to radicals being trapped at multiples of the unit cell, but only along the crystal axes. The paper notes that this is the first report of radiation-generated radicals trapped preferentially only along the three crystallographic axes. Unfortunately, these remarkable $V(T)$ were not shown. The $F(r)$ extracted from the PELDOR data lies almost entirely between 2.0 and 5.5 nm with ripples at multiples of the unit cell dimension. This distribution was reproduced very closely by detailed simulations incorporating the crystal structure and anisotropic recombination. The observed and simulated differences in $F(r)$ with different types of radiation were suggested as providing a way to determine LET of unknown radiation by the ammonium tartrate dosimeter.

These brief examples show the promise of PELDOR for future investigations of the spatial properties of paramagnetic particles participating in elementary acts of photo- and radiation chemistry. The success with photoexcited photosynthetic systems emphasizes that PELDOR is applicable to transient excited states as well as stabilized radicals. Applications of PELDOR to functioning conductive polymers and photoconversion devices appear to be feasible.

References

1. Salikhov KM, Semenov AG, Tsvetkov YD (1976) Electron spin echo and its applications. Nauka, Novosibirsk
2. Salikhov KM, Tsvetkov YD (1979) Electron Spin echo studies of spin-spin interactions in solids. In: Kevan L, Schwartz RN (eds) Time domain electron spin resonance. John Wiley, New York, pp 231–278
3. Bowman MK, Maryasov AG, Tsvetkov YD (2014) EPR measurement of the spatial distribution of radiation damage. In: Lund A, Shiotani M (eds) Applications of EPR in radiation research. Springer, Heidelberg, pp 581–627. https://doi.org/10.1007/978-3-319-09216-4
4. Milov AD, Salikhov KM, Shirov MD (1981) Application of eldor in electron-spin echo for paramagnetic center space distribution in solids. Fiz Tverd Tela 23(4):975–982
5. Bagdasaryan KS (1976) Two-quantum photochemistry. Nauka, Moscow
6. Milov AD, Ponomarev AB, Tsvetkov YD (1984) Electron electron double-resonance in electron-spin echo—model biradical systems and the sensitized photolysis of decalin. Chem Phys Lett 110(1):67–72. https://doi.org/10.1016/0009-2614(84)80148-7
7. Klette R, Torring JT, Plato M, Mobius K, Bonigk B, Lubitz W (1993) Determination of the *g* tensor of the primary donor cation radical in single-crystals of *Rhodobacter sphaeroides* R-26 reaction centers by 3-mm high-field EPR. J Phys Chem-Us 97(9):2015–2020. https://doi.org/10.1021/j100111a047
8. Hoff AJ (ed) (1989) Advanced EPR: applications in biology and biochemistry. Elsevier, Amsterdam. https://doi.org/10.1016/b978-0-444-88050-5.50004-5

9. Kawamori A, Mino H, Hara H, Astashkin AV, Tsvetkov YD (1995) Multiple resonance in pulsed EPR with application to photosynthesis. Kwansei Gakuin Univ Annu Stud 64:221–231
10. Kawamori A (2003) Electron transfer and structure of plant photosystem II. In: Lund A, Shiotani M (eds) EPR of free radicals in solids: trends in methods and applications. Springer US, pp 529–563. https://doi.org/10.1007/978-1-4757-5166-6
11. Hara H, Kawamori A, Astashkin AV, Ono T (1996) The distances from tyrosine D to redox-active components on the donor side of Photosystem II determined by pulsed electron-electron double resonance. Bba-Bioenergetics 1276(2):140–146. https://doi.org/10.1016/0005-2728(96)00071-0
12. Kawamori A, Katsuta N, Arao S, Ishii A, Minagawa J, Mino H, Ono T (2001) Three-dimensional structure of photosystem II studied by pulsed EPR. In: Proceedings of 12th international congress on photosynthesis. CSIRO Publishing, Collingwood, Victoria
13. Kawamori A, Ono TA, Ishii A, Nakazawa S, Hara H, Tomo T, Minagawa J, Bittl R, Dzuba SA (2005) The functional sites of chlorophylls in D1 and D2 subunits of Photosystem II identified by pulsed EPR. Photosynth Res 84(1–3):187–192. https://doi.org/10.1007/s11120-005-1000-y
14. Kawamori A, Katsuta N, Hara H (2003) Structural analysis of three-spin systems of photosystem II by PELDOR. Appl Magn Reson 23(3–4):557–569. https://doi.org/10.1007/Bf03166641
15. Kawamori A, Katsuta N, Mino H, Ishii A, Minagawa J, Ono TA (2002) Positions of Q_A and Chl_Z relative to tyrosine Y_Z and Y_D in photosystem II studied by pulsed EPR. J Biol Phys 28 (3):413–426. https://doi.org/10.1023/A:1020312621949
16. Kuroiwa S, Tonaka M, Kawamori A, Akabori K (2000) The position of cytochrome b_{559} relative to Q_A in photosystem II studied by electron-electron double resonance (ELDOR). Biochim Biophys Acta 1460(2–3):330–337
17. Astashkin AV, Hara H, Kawamori A (1998) The pulsed electron-electron double resonance and "2+1" electron spin echo study of the oriented oxygen-evolving and Mn-depleted preparations of photosystem II. J Chem Phys 108(9):3805–3812. https://doi.org/10.1063/1.475770
18. Jeschke G, Bender A, Schweikardt T, Panek G, Decker H, Paulsen H (2005) Localization of the N-terminal domain in light-harvesting chlorophyll a/b protein by EPR measurements. J Biol Chem 280(19):18623–18630. https://doi.org/10.1074/jbc.M501171200
19. Savitsky A, Dubinskii AA, Flores M, Lubitz W, Mobius K (2007) Orientation-resolving pulsed electron dipolar high-field EPR spectroscopy on disordered solids: I. structure of spin-correlated radical pairs in bacterial photosynthetic reaction centers. J Phys Chem B 111 (22):6245–6262. https://doi.org/10.1021/jp070016c
20. Schnegg A, Dubinskii AA, Fuchs MR, Grishin YA, Kirilina EP, Lubitz W, Plato M, Savitsky A, Mobius K (2007) High-field EPR, ENDOR and ELDOR on bacterial photosynthetic reaction centers. Appl Magn Reson 31(1–2):59–98. https://doi.org/10.1007/Bf03166248
21. Mobius K, Lubitz W, Savitsky A (2013) High-field EPR on membrane proteins—Crossing the gap to NMR. Prog Nucl Mag Res Sp 75:1–49. https://doi.org/10.1016/j.pnmrs.2013.07.002
22. Fuchs MR, Schnegg A, Plato M, Schulz C, Muh F, Lubitz W, Mobius K (2003) The primary donor cation $P^{+\cdot}$ in photosynthetic reaction centers of site-directed mutants of *Rhodobacter sphaeroides*: g-tensor shifts revealed by high-field EPR at 360 GHz/12.8 T. Chem Phys 294 (3):371–384. https://doi.org/10.1016/s0301-0104(03)00319-7
23. Bowman MK, Budil DE, Closs GL, Kostka AG, Wraight CA, Norris JR (1981) Magnetic resonance spectroscopy of the primary state, P^F, of bacterial photosynthesis. P Natl Acad Sci USA 78(6):3305–3307
24. Norris JR, Bowman MK, Budil DE, Tang J, Wraight CA, Closs GL (1982) Magnetic characterization of the primary state of bacterial photosynthesis. Proc Natl Acad Sci U S A 79 (18):5532–5536

25. Closs GL, Forbes MDE, Norris JR (1987) Spin-polarized electron-paramagnetic resonance-spectra of radical pairs in micelles—observation of electron-spin spin interactions. J Phys Chem-Us 91(13):3592–3599. https://doi.org/10.1021/j100297a026
26. Stehlik D, vanderEst A, Kamlowski A (1996) Transient spin states in the primary processes of photosynthesis. Ber Bunsen Phys Chem 100 (12):2028–2035
27. Hasharoni K, Levanon H, Tang J, Bowman MK, Norris JR, Gust D, Moore TA, Moore AL (1990) Singlet photochemistry in model photosynthesis—identification of charge separated intermediates by fourier-transform and CW-EPR spectroscopies. J Am Chem Soc 112 (18):6477–6481. https://doi.org/10.1021/ja00174a004
28. Bowman MK, Chen HJ, Maryasov AG (2017) Pulsed EPR Signals from Triplets. Z Fur Phys Chem-Int J Res in Phys Chem Chem Phys 231(3):637–652. https://doi.org/10.1515/zpch-2016-0869
29. Schweiger A, Jeschke G (2001) Principles of pulse electron paramagnetic resonance. Oxford University Press, Oxford, UK, New York
30. Stowell MHB, McPhillips TM, Rees DC, Soltis SM, Abresch E, Feher G (1997) Light-induced structural changes in photosynthetic reaction center: Implications for mechanism of electron-proton transfer. Science 276(5313):812–816. https://doi.org/10.1126/science.276.5313.812
31. Savitsky A, Niklas J, Golbeck JH, Mobius K, Lubitz W (2013) Orientation resolving dipolar high-field EPR spectroscopy on disordered solids: II. Structure of spin-correlated radical pairs in photosystem I. J Phys Chem B 117 (38):11184–11199. https://doi.org/10.1021/jp401573z
32. Tsvetkov YD (1983) Study of spatial peculiarities of formation and reactions of radicals in a solid-phase by the electron-spin echo method. Usp Khim 52(9):1514–1537
33. Tsvetkov YD (1989) ELDOR in ESE study of magnetic dipole-dipole interactions. In: Keijzers CP, Reijerse EJ, Schmidt J (eds) Pulsed EPR: a new field of applications. North Holland, Amsterdam, pp 206–218
34. Ponomarev AB, Milov AD, Tsvetkov YD (1988) Double electron-electron resonance in electron-spin echo and the spatial distribution of radicals formed by irradiation of frozen cyclohexane. Khim Fiz 7(12):1673–1679
35. Ponomarev AB, Milov AD, Tsvetkov YD (1990) Double electron-electron resonance in electron-spin echo—spatial distribution of radicals forming during radiolysis of polyethylene, monocarboxylic and dicarboxylic-acids. Khim Fiz 9(4):498–503
36. Bowman MK, Becker D, Sevilla MD, Zimbrick JD (2005) Track structure in DNA irradiated with heavy ions. Radiat Res 163(4):447–454. https://doi.org/10.1667/rr3338
37. Marrale M, Brai M, Barbon A, Brustolon M (2009) Analysis of the spatial distribution of free radicals in ammonium tartrate by pulse EPR techniques. Radiat Res 171(3):349–359. https://doi.org/10.1667/Rr1358.1
38. Jeschke G (2012) DEER distance measurements on proteins. Annu Rev Phys Chem 63 (1):419–446. https://doi.org/10.1146/annurev-physchem-032511-143716
39. Tsvetkov YD, Milov AD, Maryasov AG (2008) Pulsed electron–electron double resonance (PELDOR) as EPR spectroscopy in nanometre range. Russ Chem Rev 77:487–520

Chapter 9
Conclusions

9.1 PELDOR and Other Distance Measured Methods

In closing, let us briefly compare the PELDOR technique to other structural methods to define its place among them and to summarize the relative advantages and disadvantages.

X-ray crystallography is the most widely used method for determining inter-atomic distances. It requires good single crystals of the material. In the case of biomolecules, obtaining single crystals may be problematic through lack of the material or instability, poor solubility, or an inability to find the proper conditions for crystallization. As a rule, these problems are difficult to overcome if intricate biological complexes, e.g., protein–membrane complexes, must be analyzed. A less obvious limitation is that the conformations that crystallize are not always the biologically-relevant conformations. The X-ray diffraction process damages the molecules by breaking chemical bonds and can also change the redox state of redox-active centers. The structure obtained from a crystal may not be the structure of the biomolecule in solution, in a cell, or even the intact biomolecule.

EPR methods and, in particular, PELDOR are free from these limitations. A few picomoles of the biomolecules are sufficient for analysis; they can be investigated in various molecular environments in defined redox and pH states, and in complexes with other biosystems. The most important feature of the PELDOR is the fact that this method offers a chance to analyze chaotically-oriented samples. Even intrinsically-disordered proteins or polymers can be studied.

In chaotically-oriented systems, distances can be determined using the fluorescence resonant energy transfer method (FRET) [1]. This optical method is based on the quantum yield of fluorescence from energy transfer between a donor and an acceptor chromophore, which usually must be introduced in the molecule as labels. The mechanism of this transfer can be attributed to dipole–dipole interaction between the electric transition dipoles of the chromophores. The transfer efficiency is proportional to $\alpha/r^{6,}$ where α is a function defined in auxiliary experiments. This

© Springer Nature Switzerland AG 2019
Y. D. Tsvetkov et al., *Pulsed Electron–Electron Double Resonance*,
https://doi.org/10.1007/978-3-030-05372-7_9

method has enjoyed wide application thanks to its high sensitivity, as little as one molecule can be detected in an experiment, and the possibility of making measurements in liquid phase. The range of measurable distances almost coincides with the PELDOR range. Among the drawbacks of FRET are the necessity to independently determine the function α, and the size of the chromophores, which are more rigid and bulkier than most spin labels. These drawbacks lead to considerable problems in interpreting distances, to a lower accuracy for r, and to the impossibility of determining $F(r)$.

NMR methods have been widely used to measure distances of a few nanometers. However, NMR has several limitations. One is that only rather short distances, fractions of a nanometer, can be measured directly. Larger distances are obtained from molecular models that try to satisfy all the measured short-range distances. Portions of a molecule having many coexisting conformations may be uncharacterized. Owing to the smaller magnetic moments of nuclei in comparison with the electron magnetic moment, the range of measurable distances in solid state NMR is usually limited to a few nanometers. However, versatility, the highly sophisticated instrumentation and supporting software tools, and broad availability are undeniable advantages of the NMR methods. NMR also benefits from highly developed methods for labeling biomolecules with specific isotopic labels.

The simplest EPR method for measuring distances by dipole broadening is to analyze the width and shape of the CW EPR spectrum. Methods for simulating the spectrum shape and mathematical deconvolution have been developed and reviewed in [2]. Along with dipole interactions, HFI, exchange and quadrupole interactions that make additional contributions to the inhomogeneous linewidth can also be included in the simulation. Such simulations are typically applied for distances of approximately 1.5–2.5 nm, since, at larger distances, the contribution from the dipole interactions becomes negligible compared to other sources of broadening, and simulations fail to obtain reliable and consistent values for the dipole coupling.

The so-called half-field EPR method is applicable for small distances. The forbidden transition between the levels with $m_s = -1$ and $m_s = +1$ for a spin pair in a triplet state become weakly allowed for spins closer than $r \sim 0.5$ nm. This transition has $g \sim 4$ and lies at a magnetic field $\sim H_0/2$, where H_0 is the resonant EPR field of the main allowed transitions. The intensities $I_{\pm 1}$ of these forbidden lines are quite low compared to the intensity I_0 of the allowed lines; nevertheless, the distance between spins can be determined because $I_{\pm 1}/I_0 \propto 1/r^6$ [3]. Various CW and some pulse EPR methods to determine distance are reviewed in [2].

The three and four pulse versions of PELDOR are currently the most widely applied pulsed dipole spectroscopy methods. Single frequency pulse methods, such as "2 + 1" [4] or the single frequency technique for refocusing dipolar couplings (SIFTER) [5], are used much less often. However, the single-frequency double quantum coherence (DQC) method is frequently used [6, 7]. In DQC, the EPR spectrum is fully excited by a sequence of six pulses, and the decay of one of the spin echo signals, which is modulated by the dipolar frequencies, is measured.

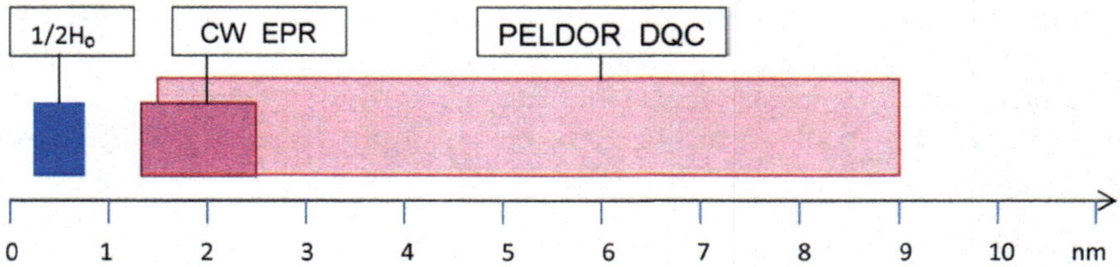

Fig. 9.1 Approximate range of distances measured by CW and pulse EPR spectroscopy

This special pulse sequence provides a way to extract the pairwise dipolar interactions as in PELDOR. DQC places stringent requirements on the hardware and requires more sophisticated theoretical treatments than does PELDOR. Investigations into the composition and properties of various biomolecular structures using the DQC method are reviewed in detail in [8]. As noted above, DQC is efficient in measuring distances up to $r \sim 7.0$–8.0 nm and has greater sensitivity because the full EPR spectrum is excited and contributes to the signal instead of just a portion, as is the case of PELDOR. On the other hand, DQC does not seem capable of giving information about relative orientations of the spins that are available from orientation selection in PELDOR. Like the other single-frequency pulse techniques, DQC is used only in a few laboratories. The approximate range of distances measured by CW and pulse EPR spectroscopy are shown in Fig. 9.1.

The PELDOR technique holds an important place among other structurally-oriented methods of radiospectroscopy, such as CW and pulse EPR, NMR, and NQR. The simplicity of the measurement and the modest instrumental requirements make it available for many laboratories, together with the established methods and software for interpreting its results, are important features of the PELDOR technique. The necessity to introduce paramagnetic centers, e.g., spin labels, is partially compensated for by the large arsenal of special methods, e.g., SDSL, for precisely introducing labels into simple molecules as well as complex biological structures.

9.2 Summary

In conclusion, let us review the main features of the PELDOR technique.

- PELDOR eliminates the inhomogeneous broadening in EPR spectral lines and obtains a direct measurement of the magnetic dipole–dipole interactions in non-oriented systems;
- PELDOR routinely measures distances between paramagnetic centers with a high accuracy in the range of $\sim$ 1.5–8.0 nm, and in some systems, to even twice that distance;

- PELDOR provides the distance distribution function $F(r)$ from the time trace $V(t)$ and can determine other features of the spatial distribution of paramagnetic centers, e.g., their mutual orientation; and
- PELDOR provides an estimation of the number N of centers coupled by dipolar interactions in spatially distinct groups, i.e., complexes, aggregates, clusters, etc.

The PELDOR method is available to a wide circle of researchers having access to pulse EPR spectrometers because of its modest hardware requirements. The theory supporting PELDOR has been developed and is expressed in software tools that support analysis of experimental results to obtain structural information.

PELDOR spectroscopy allows one to go beyond the measurement of distances. One can begin to investigate aggregation; the formation of supramolecular complexes; the interaction of various biologically-important structures with membranes; and even dynamic processes involving paramagnetic particles. PELDOR spectroscopy, together with other high-resolution EPR methods, has a rich future with many more applications in physics, chemistry, and biology.

References

1. Lakowicz JR (2006) Principles of fluorescence spectroscopy. Springer US. https://doi.org/10.1007/978-0-387-46312-4
2. Berliner LJ, Eaton GR, Eaton SS (2002) Distance measurements in biological systems by EPR. biological magnetic resonance, vol 19. Springer, Boston. https://doi.org/10.1007/b111467
3. de Groot MS, van der Waals JH (1960) Paramagnetic resonance in phosphorescent aromatic hydrocarbons. Mol Phys 3(2):190–200. https://doi.org/10.1080/00268976000100221
4. Kurshev VV, Raitsimring AM, Tsvetkov YD (1989) Selection of dipolar interaction by the 2+1 pulse train ESE. J Magn Reson 81(3):441–454. https://doi.org/10.1016/0022-2364(89)90080-2
5. Jeschke G, Pannier M, Godt A, Spiess HW (2000) Dipolar spectroscopy and spin alignment in electron paramagnetic resonance. Chem Phys Lett 331(2–4):243–252
6. Borbat PP, Freed JH (2002) Double-quantum ESR and distance measurements. In: Berliner LJ, Eaton SS, Eaton GR (eds) Distance measurements in biological systems by EPR. vol 19. Springer, New York. https://doi.org/10.1007/b111467
7. Borbat PP, Freed JH (1999) Multiple-quantum ESR and distance measurements. Chem Phys Lett 313(1–2):145–154. https://doi.org/10.1016/S0009-2614(99)00972-0
8. Borbat PP, Freed JH (2007) Pros and cons of pulse dipolar spectroscopy: DQC and DEER. EPR Newsletter, 2007

Printed by Printforce, the Netherlands